THE SPIKE

THE SPIKE

THE SPIKE

AN EPIC JOURNEY
THROUGH THE BRAIN
IN 2.1 SECONDS

MARK HUMPHRIES

PRINCETON UNIVERSITY PRESS
PRINCETON AND OXFORD

Published by Princeton University Press
41 William Street, Princeton, New Jersey 08540
99 Banbury Road, Oxford OX2 6JX

press.princeton.edu

First paperback printing, 2023
Paperback ISBN 978-0-691-24148-7
Cloth ISBN 978-0-691-19588-9
ISBN (e-book) 978-0-691-21351-4

British Library Cataloging-in-Publication Data is available

Editorial: Hallie Stebbins and Kristen Hop
Production Editorial: Brigitte Pelner
Cover Design: Lauren Smith
Cover image by SciePro / iStock
Production: Jacqueline Poirier
Publicity: Sara Henning-Stout (US) and Kate Farquhar-Thomson (UK)
Copyeditor: Dawn Hall

This book has been composed in Arno Pro with League Mono display

To Nic, Abbi, and Seth

CONTENTS

THE SPIKE

CHAPTER 1

Introduction

ENTER SPIKES

Midafternoon is the devil's time. The ebbing of your circadian rhythms collides with the digestion of an ill-considered lunch of hot dog and hummus to dull your mind and bring thoughts of a cheeky nap. But there's an all-hands meeting in the conference room in ten minutes, at which you've discovered that snoring loudly enough to drown out the CEO's "always be coding" speech is a no-no. Eat something, says inner you. On the desk abutting yours is the box for some homemade ginger, pear, and chocolate cookies that Dietrich brought for the 10:00 a.m. conference call with the South Africa office—strangely delicious, definitely tempting, disappointingly gone.

No, wait. Your eyes glimpse a rounded, crumbly edge. There's one left. Your brain sparks to life, as you glance around to clock your coworkers' locations and think—*could I take that?* After a moment's hesitation, weighing the ethical dilemmas and more importantly confirming that no one has line of sight, you extend a hand.

In those few moments your brain is abuzz with electricity. Vital, surreptitious-cookie-obtaining electricity. Why?

Your brain uses electricity to communicate. Each nerve cell, each of the eighty-six billion neurons in your brain, talks to other neurons by sending a tiny blip of voltage down a gossamer thin cable. We neuroscientists call that blip "the spike." These tiny pulses of electricity stream endlessly, ceaselessly across your brain. Spikes are seeing, hearing, and

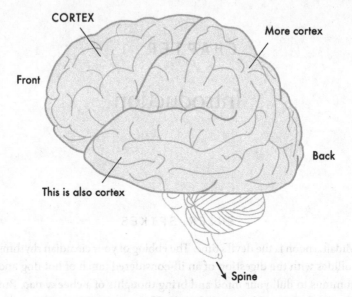

FIGURE 1.1. Basic anatomy of the human brain. Most of the outside of
your brain is the cortex.

feeling; thinking, planning, and doing. Spikes are how neurons talk to
each other. And neurons talking to each other is how you do anything.

A LIFE IN SPIKES

The uniquely human things you do are thanks to the chatter of spikes
in your cortex (figure 1.1). This outer layer of the brain contains more
neurons in you than in any other animal, ever.[1] So many in fact that we
have to divide the cortex into a constellation of areas, each with its own
name, to make sense of it all. (Few of these names are exciting—the area
with the most neurons that talk directly to the spine, and so has the
most control over movement, is called the primary motor cortex; the
areas next door are the premotor cortex and, wait for it, the supplemen-
tary motor area. Inspired.) These areas all share the same types of neurons but do wildly different things with the spikes sent between them.

Many of these areas are dedicated to seeing, from the areas breaking down the world into its simplest components—edges and lines and corners—to the areas dealing with motion, colors, objects, and faces. Some areas do hearing and touch; some control your movement.

There are areas for uniquely human things, like reading, speaking, and understanding the spoken word. And at the front of the cortex we find areas that do mysterious things with information from the outside world, somehow using it to plan, anticipate, and predict. All of it done by spikes.

The numbers are vertiginous. Of the eighty-six billion neurons in the adult human brain, about seventeen billion of those are found in the cortex. Each of those sends at most one spike per second, on average.[2]

The United Nations tells us the expected lifespan of a human on this planet is about seventy years. That's more than two billion seconds, each of which contains about seventeen billion spikes in cortex. All told, your lifespan is about thirty-four billion billion cortical spikes.

The cry you emitted on your appearance in the world. Your first tottering, uncertain steps. The pain when Susan's wildly swinging arm knocked out your wobbly tooth in primary school. Recognizing that cluster of trees in the distance, and the relief of knowing you'll now find your way across the damp, foggy hills back to the welcome warmth of the car. Plucking up the courage to ask for a date, and blurting it all out in a rush. The flush of embarrassment. The quiet euphoria of a yes. Deciding you just have to do something about the clash between the purple sofa and the lime green curtains. Remembering the smell of your mom's bread and dad's roast chicken. Cradling your baby. Reading this sentence. And this one.

All spikes.

From the magnificent to the mundane, everything you've done is in those thirty-four billion billion spikes that have streamed across your cortex. If I were to write the story of your life with one word for every spike, your biography would be longer than the combined length of all novels in English ever published.[3] Yes, ever, since Gutenberg introduced movable type to Europe in 1439. And not just a bit longer—seventy-six million times longer. Even with the combined efforts of Tom Wolfe,

Neal Stephenson, and George R. R. Martin to deliver novels that are also handy for weighting down small children in a storm, novelists still have at least another 380 million years or so to publish as many words in English as spikes in your cortex in your lifetime. And below the cortex, billions upon billions more neurons, sending billions upon billions more spikes.

You'll excuse me if I attempt something a little less daunting.

THE JOURNEY OF A SPIKE

In this book, I'm going to tell you the story of just two of those seconds. Of a simple act: you spot that last cookie in the office tray, and think—*no one will mind if I take that, right?*

A spike's journey from the eye that receives the light bouncing from the cookie, through the seeing parts of cortex turning patterns of light and shade into the edges, curves, crumbs, and colors of the cookie, on to cortical areas for perceiving, deciding, and remembering, plunging into the depths of the motor system, and out, out through the spine and on to the muscles, finally moving your hand to what your eye can see. A journey from seeing to deciding to moving, from eye to hand.

This is the story of everywhere the spike was sent, and of everything it saw on the way—the twinkling galaxy of neurons, the deep darkness of the cortex, the loneliest neuron. Of splitting into a thousand clones. Of spontaneous birth and instant death. An epic journey, all in but a moment of time, a story replayed two billion times over.

THE GOLDEN AGE

That I can tell you this story at all is thanks to a remarkable convergence of technologies.

One of these is brain imaging, especially functional MRI (fMRI). Relied on heavily in popular accounts of neuroscience, fMRI can tell us much about the broad picture, of how a group of brain areas may process vision, but not hearing; create emotional responses to faces, but not chocolate; or paradoxically only turn on when your mind goes blank. Yet fMRI tells us nothing about how neurons work. Each tiny pixel on a fMRI image, each dot of color, contains 100,000 neurons. fMRI measures the flow of oxygen-rich blood around those 100,000 neurons, a flow that increases as those 100,000 neurons send more spikes, for making spikes needs energy, and creating energy needs oxygen. Each dot of color shows us only where the demand for such energy-giving blood has changed around 100,000 neurons. So fMRI cannot see or record individual neurons, let alone the spikes emitted from them.

It is a wonderful technology, the only way to peer at the moment-to-moment activity inside the living human mind, and with great potential for our assault on neurological disorders, where diagnosis and treatment perhaps take precedence over a deep understanding of what each neuron is doing. But alone it is of no use to us here. Trying to understand how neurons work using fMRI is like trying to follow a soccer match through the roar of the crowd. The crowd's crescendos and groans will tell you when something exciting is happening, and with luck which part of the crowd is baying will tell you roughly at which end of the field the excitement is happening in. But you'll be oblivious to the match itself, to the intricacies of what the players and the ball whizzing between them have been doing for ninety minutes. To understand a match, we need to watch the players. To understand the brain, we need to watch the spikes.

We caught our first glimpse of the spike from a single neuron in the 1920s.[4] Since then, tens of thousands of neuroscientists have recorded spikes from every imaginable part of the brain. And from almost every imaginable brain, from the giant tentacle neurons of the squid, to the deciding neurons of the rat, even to the neurons in an awake, chatty, lucid human. But now we can go further, for we are in the midst of the golden age of systems neuroscience, the pursuit of how neurons are wired and work together.

For decades we could record the spikes of only one neuron at a time. Now we can record the spikes of hundreds or thousands at the same time with standard equipment, and the cutting edge is growing exponentially year on year.[5]

We used only to be able to trace the broad outlines of where neurons in one area of the brain sent their cables outward. Now we can trace the wiring of each single neuron to find out precisely where spikes will be sent.

We can now record not just the spikes coming out of a neuron but also the tiny effect they have on the next neuron, at a connection smaller than a bacterium. We can even do so at multiple sites on a single neuron at once.

More than just record them, we can now turn neurons on and off with light, either forcing them to send spikes on command, or stopping them from sending spikes altogether.[6] So we can at last test directly what spikes are for, by seeing what happens when they are sent, or, just as importantly, not sent.

Combined, these tools let us record the spikes sent from hundreds of separate neurons, stop or start spikes at will, and give us some idea of the destinations of the wires along which they travel. Combined, these tools can now tell us the journey of spikes.

There's a catch to this smorgasbord of technological triumphs. None of them can be used in humans. Tracing the wiring between neurons would mean injecting fluorescent chemicals directly into a region of your brain, then taking out your brain, slicing it up, and sticking the slices under a microscope to find out where the fluorescent chemicals ended up. Can't do that to you. To turn neurons on and off with light we have to make them sensitive to light in the first place, by inserting DNA from light-sensitive plants or bacteria into the neuron's DNA. Can't do that to you either. And to record the spikes from hundreds of neurons at the same time means either filling your neurons with a toxic chemical that glows according to how active the neuron is, or sticking tens of long tungsten or carbon-fiber electrodes through your skull and into your brain, attached to long wires. Ethically speaking, the slicing, gene-fiddling, and electrode-poking are right out.

Except in fascinating rare cases. On rare occasion, we do get to record spikes from electrodes implanted into a live human brain. Sometimes these are from patients with Parkinson's disease who are undergoing surgery for deep brain stimulation. This treatment for Parkinson's targets electrical stimulation at regions deep in the brain (hence, "deep brain stimulation"—neurologists are some of the most literal-minded people on the planet). It requires a permanently implanted electrode, attached to a battery pack installed under the collarbone. The surgery to implant the electrode happens in two stages. The electrode is inserted first, into approximately the right place, but its leads are left dangling outside the skull so that the position of the electrode can be fine-tuned. During the tuning, the neurologist will pass stimulation down these leads into the electrode, and out into brain. If the electrode is in slightly the wrong place, then slightly the wrong thing will happen: if the patient salutes you, this is wrong, move the electrode a little; if the patient starts weeping uncontrollably, this is wrong, move the electrode a little. If the patient's tremulous arm suddenly becomes still, this is right; so now the electrode can be secured in place, and the second stage of surgery commences to run the leads under the skin and down to the battery pack, and to close up the hole in the skull.

But this slow fine-tuning means there is a window of time, about a week, in which these leads hanging out of the skull can also be used to record from the electrode, record the neurons next to it.[7] Creative researchers spend this week asking the patient to do a whole bunch of tasks, which hopefully will involve the tiny deep brain structure in some way. Along similar lines, people whose severe epilepsy is not responding to drugs can also have implanted electrodes, targeting stimulation at the small region of brain—typically in the hippocampus or cortex—where the seizure activity starts. Again, while getting the electrode into position, the researchers can record from neurons next to those electrodes during tasks in these patients.[8] From both rare occurrences we get precious rare recordings of single neurons from a live human. A valuable resource, but one limited to a handful of brain regions in a handful of people—and still no slicing or gene manipulation allowed.

With humans literally off the table in the quest to understand spikes, neuroscientists gather much of their data from a wide range of nonhumans. Some are our close cousins, evolutionarily speaking—the rat and mouse, in particular, for their combination of smarts and well-studied DNA. Others are studied for the unique ways they can tell us about the fundamentals of how neurons talk to each other. Salamanders, zebrafish, leeches, sea slugs, even the maggots of vinegar flies, will all appear in the pages that follow. For neurons are extraordinarily preserved from deep evolutionary time. Neurons are recognizably neurons in practically everything with some kind of brain. If you can see it, and it moves, it lives a life of spikes.

HOW WE CAN INTERPRET SPIKES

Our interpretation of these reams of data from nonhumans, data on spikes and where and when they are sent, relies on casting it into what we know about the human brain. From brain imaging we can get confirmation that similar brain regions in humans are active, at similar times and places, in response to similar things in the world, as the spikes we record in nonhumans. From psychology and the cognitive sciences we can get an understanding of what processes are happening in the human mind when those spikes are observed in nonhumans.

The face code is a beautiful example of this interplay between psychology, brain imaging, and spikes. Humans pay a lot of attention to faces. Psychology tells us that our preference for looking at faces is there from our earliest childhood, that as adults we can remember around five thousand faces,[9] and that we can recognize faces from exceptionally impoverished information: from an extraordinary variety of angles, with just a glimpse, and using the most basic of visual clues. Even this :-o. Or this ;-). Our deep ability to process faces is perhaps not surprising when you consider that recognizing faces and their expressions is the basis of many social interactions, for knowing who is kin and who is not, who is above us and who below in the pecking order, and who is pleased to see us—and who is really not. But the depths to which our minds process faces implies our brains must dedicate some serious processing power to the face problem.

Brain imaging showed us that indeed the human brain takes this problem so seriously it dedicates a whole area just to faces. The now-named "fusiform face area" lights up in humans when shown a face, at whatever odd angle you choose, but not when shown objects or scrambled faces. It really does care only about faces.[10]

Doris Tsao, Winrich Freiwald, and colleagues then sought out some nonhumans that also care about faces—monkeys—to venture into this area of their brains, record spikes, and find out the actual messages being passed between neurons.[11] They found a mass of dedicated neurons that all sent spikes in responses to pictures of faces.[12] There turned out to be six separate patches of face neurons in this one area, and these patches were linked together. Stimulation of one patch activated neurons in some of the others,[13] which suggested a face was represented by the joint activity between neurons in different patches. That joint activity code was revealed nine years later in 2017: each neuron sends spikes in response to some abstract feature common to faces—like the curve formed by the eyebrow and nose. The combination of neurons with different abstract features sending spikes together adds up to a whole face.[14]

Psychology tells us how much humans care about faces, and how deeply they process them. Brain imaging shows us a brain region dedicated to processing faces. Spikes show us the face code—how that region sends messages about faces. Recording spikes alone in response to faces would not tell us that these spikes correspond to "seeing" faces, for "seeing" is a subjective human experience. We interpret spikes in nonhumans through our own experience as humans.

WHERE WE WILL GO

In this golden age, cutting-edge technologies have only just begun to draw back the curtain on the neuronal drama of the brain. Seemingly every day of the past ten years brought new research that upended our understanding of how neurons talk to each other. And so upended our understanding of what makes us tick—of how we see, of how we decide, of how we move. But each cluster of neuroscientists working feverishly on their favorite brain region or type of neuron cannot see the big picture, cannot know all the ways in which our understanding of the inner workings of the brain has radically changed. That's my challenge.

By taking you on the journey of a spike from eye to hand, this book will tell the story of what we know about spikes, about what they mean for us as humans, and of what we have left to understand. Journeying with the spike will let us rip apart misconceptions about how brains work and about how they fail, many of these held by neuroscientists themselves.

A textbook neuron has a defined function, a defined reason why it sends spikes, based on some external cause in the world. But we will meet the dark neurons, the literal silent majority, who sit unmoved by anything and everything going on around them. They are invisible to brain imaging and challenge our most deeply held theories of what neurons do. Evolution tolerates no waste, so why would it allow there to be billions of neurons that apparently do nothing?

And we will meet the spontaneous spikes. Spikes mysteriously created by neurons without any input from the outside world; spikes created solely by the myriad feedback loops between neurons that drive each other to spike endlessly. They carry no message from the world, or to the world through movement. Crazier still are spikes born without any input even to the neuron that created them, spikes created solely by the internal cycling of molecules inside a neuron. Yet on our journey from seeing to moving, we will meet these spikes everywhere.

Meeting the spontaneous spikes leads to one of the new ideas I will advance in this book: that spontaneous spikes are an inevitable consequence of wiring up a big bag of neurons into a brain—and evolution has co-opted them for survival. Rather than waiting for spikes to make their journey through the myriad areas of cortex to work out what is being seen, to decide what to do about it, and then to act—rather than wait for all that, we have harnessed spontaneous spikes to give us the power of anticipation. Spontaneous spikes predict what we'll see next, what we'll hear next, what our next decision is likely to be. They prepare for our next movements. All so we can react faster, move quicker—and survive longer.

Clinging to a spike as it speeds from your eye through your brain to your hand, from seeing the cookie to deciding to nab it to reaching it, we will trace torturous paths, be cloned, and fail badly. We will wander through the splendor of the richly stocked prefrontal cortex and stand in terror before the wall of noise emanating from the basal ganglia. All of this is yet to come. For we start with the thing we understand best of all: the spike itself.

CHAPTER 2

All or Nothing

BINARY

Warren McCulloch made an untenable leap of faith in the early 1940s. The kind of creative daring that only his bizarre mélange of psychiatrist-neuroscientist-philosopher-theorist[1] would attempt. The first fuzzy pictures of spikes appeared in the late 1920s and early 1930s. Tiny wobbles on an oscilloscope,[2] showing electrical pulses so small they'd be vaporized by a cough in the next room. Yet McCulloch was struck by how each spike from the same neuron looked roughly the same shape, the same size, every time it appeared. With just a handful of neurons then recorded, he made a bold prediction: that every spike from every neuron in the entire brain is all-or-nothing; the spike either appears in its predetermined shape and size, or does not appear at all.

Decades of work have shown McCulloch was right. In this chapter, we use his inspired guess to tackle the existential question: why spikes?

McCulloch turned out to be right because of the way spikes are made. Like all cells, neurons have a membrane, a skin, that surrounds them, keeping their innards within. In a neuron, this skin separates a lot of ions on the inside from a lot of ions on the outside. And the difference between the charge of the ions on the inside and outside means the neuron has a tiny voltage that flickers up and down constantly.

But when the voltage of the neuron's body reaches a tipping point, it triggers a runaway, rapid-fire sequence of holes opening and closing in the neuron's skin, ions hurtling in and out, creating the electrical pulse

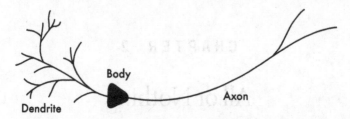

FIGURE 2.1. Important bits of a neuron. Spikes are made in the neuron's body and sent down its axon, there to arrive at the next neurons' dendrites, a tree reaching out to intercept the messages of spike-carrying axons.

that can travel far beyond the furthest reaches of the neuron's body. A spike is born, sent screaming down the neuron's axon, the cable connecting one neuron to another, to reach the next, distant target (figure 2.1).

The opening-and-closing sequence is always the same, so the spike is always the same shape and size. There is a spike or not, and nothing in between.

The journey to understanding the all-or-none nature of spikes started in easily accessible places in easily kept animals: the bullfrog's sciatic nerve, the eye of a horseshoe crab, the eye of an eel.[3] These spikes all seemed to repeat their own exact same shape every time they appeared. But starting from these recordings in the early 1930s it took over two decades of painstaking work to find the answer to why this happened, culminating when Alan Hodgkin and Andrew Huxley pulled all the evidence together in 1952.

Hodgkin and Huxley recorded from the giant squid axon (an axon that is giant in a squid; not an axon from the giant squid—getting Leviathan onto the lab table proved a trifle difficult). Because this axon was so massive, they could take it out, stick it in a bath, put an electrode inside it, and record a spike directly passing along the axon. Their clever idea was to then play with the ions in the liquid in which the neuron was sitting, increasing or decreasing the concentration of particular types of ions, in order to find out how this changes the neuron's ability to send a spike.

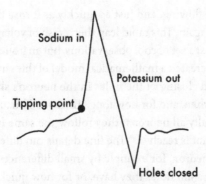

Sodium in

Potassium out

Tipping point

Holes closed

FIGURE 2.2. A spike. The neuron's voltage (thick black line) fluctuates
until it reaches the tipping point. This triggers the cascade of holes
opening in the neuron's skin, ions flooding in and then out, rapidly
driving the voltage up, then crashing back down, before returning to
normality. The whole process takes about a millisecond.

For you see, neurons sit in salty water—outside a neuron's skin are
lots of sodium (which has a positive charge: +) and lots of chlorine
(which has a negative charge: −). But inside the neuron is a little sodium,
a little chlorine, and lots of potassium (another +). Because the concentrations of each type of charged ion—particularly the potassium—are
different either side of the neuron's skin, this creates a voltage across the
skin. By playing with the concentrations of ions outside the neuron,
Hodgkin and Huxley were thus mucking about with the neuron's voltage. And crucially they could find out which types of ions (sodium and
friends) affected what part of sending a spike.

What they unpacked with their squid axon in a bath of saltwater was
the remarkable birth of a spike (figure 2.2). When the neuron's voltage
increases beyond its tipping point, suddenly holes that only permit sodium open up in the neuron's skin, and sodium ions rush in, rapidly
increasing their concentration on the inside, and voltage rockets. But
only briefly. For the onrush of sodium triggers the opening of a different
set of holes in the skin, which pump potassium back to the outside,
sending positive charge back out almost as quickly as it's arriving via the
sodium ions. In turn, this outrush of potassium shuts off the sodium

holes, ions stop flowing, and just as quickly as it rose the voltage becomes negative again. This rapid leap then crash of voltage is the spike.

This wasn't just a set of cool observations, but an ironclad law. Hodgkin and Huxley created a mathematical model of this entire process, of the opening and closing of the holes in the neuron's skin—of which holes opened, when, and for how long. Their set of mathematical laws applies to virtually all neurons: they follow the same laws each time their tipping point is reached.[4] The fine details can differ between different types of neuron, for example by small differences in how many sodium or potassium holes they have, or for how quickly those holes open and close. So the spike in a squid's giant axon can have a different shape from the spike of a neuron in a dormouse's hippocampus. But regardless of their subtle differences they are always spikes, always all-or-nothing.

By his leap of faith that this all-or-nothing electrical pulse was true always and everywhere, McCulloch realized we could radically simplify how we think about the brain. Instead of worrying about the details of the shape of the pulse, or its width, or its sloppiness, we only need to know that it was sent, or not sent. That a spike means "1" and the lack of spike means "0." That all messages sent from a neuron are binary.

And binary implies logic. McCulloch knew this much, but couldn't do the math alone. Fortune then placed in his path the otherworldly genius Walter Pitts,[5] a man who at age twelve corresponded with Bertrand Russell on errors in Russell and Whitehead's monumental *Principia Mathematica* after having read it while hiding out from bullies in a public library; ran away to the University Chicago at fourteen, working menial jobs and sneaking into lectures on math and logic (if you're thinking this sounds strangely like the plot of *Good Will Hunting*, you're not alone); and whose one good friend, Jerry Lettvin, happened to know Warren McCulloch—and to know that McCulloch needed the help of a tall, awkward, high-foreheaded, otherworldly logician genius.

Together McCulloch and Pitts proved a deep theory that a group of neurons sending 1s or 0s to each other could produce all of logic. That, for example, a pair of neurons could compute AND: by both sending a spike—a 1—if both received an input, and neither sending a spike—a o—for any other combination of their inputs. A different pair could compute OR: by each sending a spike (1) when that neuron received an input, but not sending a spike (o) if both neurons received no inputs or an input at the same time. Adding more and more neurons, McCulloch and Pitts showed, could compute all such statements of logic, no matter how convoluted. And anything that can produce all the statements of logic can compute. So it seemed that the answer to "why spikes" was: so the brain can compute.

If you know anything about digital computers—the box on your desk, the laptop on your knee, the tablet in your hand, the phone in your pocket—you may be thinking at this point: ah! Binary! So the brain is a computer! But that comparison has its history backward; rather, the digital computer is a brain.

John von Neumann laid out the architecture for modern electronic computer hardware in 1945.[6] Von Neumann knew McCulloch well, and read McCulloch and Pitts's paper; he then used the ideas of encoding os and 1s in elements of a circuit, and of how to combine these elements to do logic, in his architecture for a computer. Indeed, throughout his report laying out the EDVAC computer's architecture, von Neumann talks of his computer as being modeled on how the brain works. Computer hardware has some foundations in brain science, not the other way around.

NO LOGIC HERE

To you, sitting at your desk, suffering the drowsiness of midafternoon, craving a snack, the answer to "why spikes?" may seem more prosaic: to get me food. On the desk across from yours sits Dietrich's cardboard box, its hinged lid half-open and half-turned-away from you, revealing mostly the upside-down felt-tip scrawl of "Cookies" in a childish hand on the lid top, obscuring much, but not quite all, of the box tray below.

As your eye wanders over this scene and falls on the last lonely, tempting cookie in the tray, the light from it pours into your eye, crashes into your retina, and excites the first neurons there. Here we find something startling. The first two layers of neurons in the eye don't use spikes to talk to each other. They talk directly, constantly, in flickers of voltage and diffusions of chemicals.

Light—photons bouncing off desk and box and cookie—hits the cones at the back of your eye, the first layer of neurons in your retina. Frankly, cones are a bit weird. In darkness they are constantly releasing a stream of molecules onto the neurons of the second layer: these apparent light detectors are constantly sending messages about the absence of light. When a cone absorbs photons, its voltage drops briefly, and the steady stream of molecules pauses for a moment. The second layer of neurons, the bipolar cells, read out this pause and convert it into changes of their voltage. Some bipolar cells prefer darkness, so they read out this pause as a drop in their voltage; other bipolar cells want there to be light, so they read out this pause as an increase in their voltage. These first two layers of neurons turn light into voltage using chemistry, with nary a spike to be seen.

The second layer passes the message on to the third layer of neurons. They do this by inverting the same trick. Bipolar cells in the second layer constantly release molecules onto the neurons in the third layer, but this time the number released is proportional to the voltage of the bipolar neuron—the higher the voltage, the greater the release. In turn, receiving these molecules proportionally changes the voltage of the neurons in the third layer. The transmission from the second to the third layer turns voltage into chemicals and back into voltage again. Many of the neurons in this third layer are the ganglion cells, the neurons that talk to the rest of the brain. And to do that these ganglion cells turn their voltage into all-or-none spikes.

Clearly the retina is no mere passive collector of light, but a complex minibrain, concocted from a rich cast of characters.[7] The roll call includes three types of cones (in humans), corresponding to the three wavelengths of light that we call red, green, and blue, and the rods that let you see in the dark, which far outnumber the cones. At least nine types of bipolar cells in the second layer, plus the intricate web formed by an internal neuron that controls the flow of molecules from cones to the second layer, and more than forty types of amacrine neurons in the third layer, whose job is to control the flow of molecules from the second to the third layer. Of these fifty-plus types of neurons across both layers, the vast majority do not use spikes to send messages.

(And this lack of spikes in the eye means the neurons there cannot be doing the logical operations beloved of McCulloch and Pitts. When the first solid evidence for this absence of binary logic was produced in the 1950s, by his friends at MIT no less, Pitts burned his PhD thesis in disgust—it was, after all, on the logic of the brain.[8])

If so many neurons in the retina don't need to use spikes, why does any neuron send spikes? Why convert the flexible, continuous, analogue signal of molecules and voltage into a rigid, discrete, binary one— why seemingly throw away useful information?

The answer is that spikes let neurons send information accurately, fast, and far.

ACCURATE, FAST, AND FAR
Accurately

A spike is a time stamp that says "a thing happened just now." That thing might be the fractional change in the light falling on a frog's retina from the tiny motion of a small, curved, black object. It might be the sudden ping of the microwave oven announcing the remnants of last night's curry are reheated. It might be the sudden ramping up of pressure on the side of your tongue as you absentmindedly crush it between your molars. The thing that happened is almost certainly a change in the spikes coming from other neurons into the neuron at hand—a torrid tale taken up in the next chapter.

A spike takes less than a millisecond to make; so the timing of a spike can be accurate to less than a millisecond. Spikes are then messages that time-stamp events in the world with extreme precision.

A beautiful example of this extreme precision is how the rat's brain receives information from the rat's whiskers. The whisker system of rodents is a favorite of neuroscientists trying to understand how brains deal with sensory information because it has so few parts to deal with.

A rat has just thirty to thirty-five main whiskers on each side,[9] arranged in five neat rows, compared to more than six million cones in a human eye. We can follow the path from the nerve at the base of the whisker into the brain and track exactly which neurons respond to which whisker. Having found the very first neurons that get input from one specific whisker, we can ping that whisker and see how the neuron responds.

Rasmus Petersen's lab at the University of Manchester, with Michael Bale leading the experiments, did just this in 2015 to find out how accurately each of these first whisker neurons could send spike messages.[10] They used a tiny motor to wobble a whisker back and forth rapidly in a random pattern, and repeated that same pattern over and over again while recording from one of the neurons connected to the base of that whisker. Each pass of the whisker-wobbling pattern caused the neuron to send a stuttering sequence of spikes. If stuttering spikes were sending messages about particular changes in the whisker—perhaps how fast it is moving, or how much it is bending—then the sequence of spikes should repeat pretty closely on each pass of that pattern.

It turned out the sequence of spikes repeated so precisely that the Petersen lab ran up against the limits of even our most high-tech recording kit. This is the digital age, so the machine recording from the electrode stuck next to the whisker neuron was sampling at 24.4 kHz—it took a reading 24,400 times every second. Even with this absurdly fine-grained resolution of time there seemed to be some spikes in the sequence that occurred at the exact same moment every time they replayed the pattern of whisker-wobbles. The "exact same moment" meaning the spikes happened within the same single sample made by the machine, happened within 41 microseconds of each other. An absurdly tiny amount of time: if on the first pass of the pattern the spike occurred at, say, 3.68092 seconds, then on many of the other passes a spike also occurred somewhere between 3.68091 seconds and 3.68092 seconds. Having run up against the very limits of their technology, the scientists in the Petersen lab had to make a custom recording machine to sample at a much higher rate—500 kHz, reading from the electrode 500,000 times a second—to find out exactly how precisely the spikes were repeating.

They used this new recording machine to find the absolute limit of how precisely the spikes could be sent. They videoed the rats using their whiskers to find the fastest possible movement the whiskers ever made, as the faster the movement the more accurate the spikes it evoked would need to be. Using their tiny motor to now repeatedly move the whisker in that single ultrarapid movement over and over again, they recorded the time it took for the first spike to be sent each time the movement started. Astonishingly, the most accurate neuron repeated that first spike within about five microseconds on each pass. Thanks to spikes, a rat's whiskers can tell the rat's brain what just happened to them with extreme precision.

That spikes about rats' whiskers are highly precise is not happenstance. Whiskers are vital for rats.[11] They forage in dim and dark conditions, for which acute eyesight would be useless, and indeed a rat's eyes are a bit crap: the main job of a rat's eyes is not to give a detailed breakdown of everything in the world, but to tell the rat whether something should be approached or run away from. Whiskers are how rats find stuff, and tell what it is. Their whiskers wave back and forth constantly, around eight times per second, finding walls and seeking objects. Put a Lego brick in front of a rat, and it could not tell you what color it is. But it will explore it thoroughly with its whiskers, bending them down to the brick, and dabbing at it to understand its shape and texture.[12] A rat's whiskers are in effect its version of our eyes; so much so that when a rat really wants to study something, it will "stare" at it with its whiskers: the rat folds its whiskers forward onto the object and then waves them at up to four times the normal rate.[13] Lucky, then, that the spikes transmitted from the rat's whiskers to its brain are so accurate.

Fast

Fast things happening in the world require information about those things to be transmitted quickly into, around, and out of the brain. Ping a rat's whisker, and it'll immediately turn its head. Your glance around the office alights on the crumbly cookie, and you need to make a snap decision to pinch it. Spikes are the brain's solution to the problem of sending information quickly.

Almost all neurons in your brain have a single axon sprouting from them, a specialized cable that conveys the neuron's spikes onward to their destinations. Some axons are custom built for speed. A spike can travel about 200 millimeters per second along an axon in the cortex, covering the distance from the back to the front of your cortex in less than a second.[14] Sensory axons in the spine are a hundredfold faster still:[15] the sciatic nerve in the shrew sends spikes at 42 meters per second; in the elephant, at 70 meters per second. Or 156 miles per hour. Elephants have Ferrari nerves.

Sending information between neurons any other way is much slower. Spikes are twentyfold faster than if neurons relied instead on just spreading their voltage alone, and a thousandfold faster than trying to send messages by releasing molecules.[16] Sending these kinds of continuous messages between a pair of neurons would need them to be pressed together, touching skins, so that the slowness of the signal is compensated by the shortness of the distance to send it. This works in the first layers of your retina, where the bipolar neurons are squashed right up against the cones. But there are around seven hundred neurons covering the distance from the back to the front of your cortex.[17] So sending messages by this bucket brigade of one neighbor to the next would take impossibly long. Worse, with each passing of the message, there is a chance it will be degraded or corrupted by noise, so hundreds of such passings would ruin any message, turning "there is a cookie in the tray" to "her a Wookie ray"—leaving you stumbling hungry, and not a little confused, into your meeting. Sending spikes along fast axons circumvents all these problems.

Speed is another reason why a rat's whiskers send information to the rat's brain using spikes. When running in the dark, a rat's whiskers touch the ground ahead to make sure the way is clear—avoiding potholes, bumps, and other rodents. Rats run quickly, so on each stride their front paw will land on the same spot the whiskers were touching about 200 milliseconds ago.[18] Which means the rat's brain has less than 200 milliseconds to take in information from the whiskers, make sense of it, and react: adjust their paws and limbs, leap, or come to a screeching halt.

Spikes let the whiskers update the rat's brain, and then the rat's feet, accurately and quickly.

Far

Big bodies—and big on the scale of neurons means anything visible to the naked eye, like maggots—need their neurons to send messages over distances far greater than the size of a single neuron. Like the distance of the nerve from your fingertip to your spine, for the temperature and pressure sensors in your fingertip to tell your brain it's just put your finger in something cold, slimy, and icky and could it stop now please it feels like a slug it is a slug—urgh. Spikes are the brain's solution to the distance problem too.

A spike can travel along meters of axon. Axons connecting nearby neurons are thin; those connecting distant neurons are fat. The longer the axon, the thicker it is, and the faster the spike travels. Many far-reaching axons are covered in evenly spaced sheaths of fatty goop, myelin, that insulate the axon. This has two jobs: it allows the spike to travel fast through the goop-covered bits, and in the gap between the goop sits another set of those same holes in the skin that repeat the open-and-close cycle, regenerating the spike. They are relay stations, boosting the signal to make sure it gets to the end intact.

Sending messages between distant neurons any other way is doomed to failure. Releasing molecules can send information over tiny gaps, as we saw in the retina, and we'll see again in the next chapter. But the molecules newly released into the ocean of saltwater that surrounds neurons would get rapidly lost, as they diffuse away from where they were released; so releasing molecules is useless beyond a few micrometers. The neuron's voltage alone decays rapidly with distance, so it would become indistinguishable from noise within 1 to 2 millimeters. But sending a spike down an axon can let a neuron send a signal farther than 100,000 times its own body length. If the neuron connecting a giraffe's spine to the muscles of its back foot were the size of the Earth, its axon would stretch past the Sun.[19]

The Giraffe

Giraffes are ridiculous animals. That they are a viable animal at all is entirely thanks to spikes sending information accurately, fast, and far. Those absurdly long necks mean their brains are up to five and a half meters from their feet (OK, hooves). This presents a rather extreme control problem. When a giraffe is gamboling across the open savanna, how does it not crumple into an undignified heap every time its hoof clips a rock or branch or sleeping painted dog minding its own business? Its brain needs to react to all this.

For a giraffe not to stumble into a heap, minimally a message needs to be passed from the touch sensors in its hoof to the spinal cord, integrated with messages coming down from the brain, and then together they correct the control of the giraffe's gait by changing the signal being sent to the leg muscles from the motor neurons. So when the giraffe clips its hoof, spikes from multiple sensory neurons are sent immediately and together. The sensory axon connecting a giraffe's hoof to its spinal cord sends spikes at over 50 meters per second. The axon carrying the signal from the spine to the leg muscles is just as fast. And a single cable carries it those long distances, without tens of time-wasting stops along the way.

Accurate, fast, and far: so that when a giraffe stubs its toe, it can pull its hoof back and adjust its gait within tens of milliseconds despite the reflex neurons in its spine being meters from the ends of its legs.

EYE TO BRAIN

The need to be accurate, fast, and far is why the eye sends spikes. To get from your eye to your brain, information must cross a vast distance, from the back of the eye to a way station in the middle of the brain, a distance more than 250,000 times farther than the distance chemicals cross between neurons within your retina. A distance that only spikes can travel. And that information must get to the brain quickly and accurately, so that the ball flying toward your face can be parried, that the glass tipping from the edge of the table can be caught, that the glimpse of stripy orange fur in the tall grass just ahead of you can be identified as not the overweight tabby cat you originally took it for, nor a bloke on his way to a Winnie-the-Pooh-themed party, but a hungry, prowling tiger—and you can run. The eye turns its number crunching of the cookie into spikes and blasts them out into the vastness of your cortex.[20] Millions of them per second.

What the eye tells the brain is a complex detailed breakdown of everything that is out there. Far from just collecting light and turning it into spikes, the retina has already done much to collate, simplify, and process the image of the world.

We know much about what the ganglion cells in the third layer of the retina are telling the brain. The most basic information is "where." Light reflected from the cookie edge falls on cones at a particular location in the retina; light from the chunky chocolate chip just next to the edge will fall on cones just next door. Meaning that the activity of the cones encodes the location of the sources of light in the outside world. And this location information is preserved across the layers of neurons in the retina—cones close together connect to neurons close together in the second layer, which connect to close together ganglion cells in the third layer. Which all means that the spikes from the ganglion cells automatically tell the brain about location. (Admittedly that location is upside down and mirror-imaged with respect to the world, because light travels in straight lines through the pinhole lens in your pupil, so light from the bottom of what you can see hits the top of the retina, and vice versa; and light from the left hits the right of the retina, and vice versa.) Each ganglion cell is responsible for sending spikes about what is happening to the light in a particular location in the world.

Those spikes say that the box tray is under the cookie; the desk is under the box; and the box lid is hanging at an angle above the cookie. Well, eventually they will mean that—but not when they are sent. When they are sent, they just mean "there is a different pattern of light in this location, and this location above it and to the right, and all along this direction in a straight line." Your eye knows nothing of cookies, boxes, and desks. Your brain will work out all of that later, when the patterns of light are brought together into objects, their names are dug out of storage, and their meaning becomes apparent. The eye just knows about light, where it is, and the patterns it makes.

But there is much the retina has to say about the patterns. The next most fundamental bit of information sent by the ganglion cells is when light is on or off at a particular location. This job is split between three types of ganglion cell—ON, OFF, and ON-OFF. The ON type send spikes when the light increases in the part of the world they are responsible for. About the same number of cells are the OFF type, who send spikes when the light decreases in their location. The rarer ON-OFF type send spikes to both increases and decreases of light in their location.

While investigating what the frog's eye tells the frog's brain, Jerry Lettvin—Walter Pitts's friend at MIT—helped prove these three types existed and showed that the frog's retina had at least one more type of ganglion cell.[21] This was the "convex detector," a type of cell that sends spikes when there is light from something standing out against the background that is small and curved and moving. Or, as Lettvin and his colleagues speculated in their 1959 paper, a bug detector.

If the first three types were not enough (and they were), this "bug detector" was the killer blow to McCulloch and Pitts's ideal of the purely logical brain. For even in this first bit of brain, way out in the eye, spikes are sending messages tuned to things of relevance to the animal, tuned to its ecological niche, driven by evolution. And those spikes must be the result of a lot of processing of the light by the retina's own neurons, pooling information as they put together that the locations of light being on and light being off fall on a curve. A lot of computation, but no logic.

It gets worse. We now know that these three basic types of ON, OFF, and ON-OFF ganglion cells are themselves umbrella terms for a menagerie of neurons that actually care about vastly different things. Tom Baden, Philipp Berens, Thomas Euler, and their colleagues recently updated Lettvin's study of the frog's eye for the twenty-first century by asking what the mouse's eye tells the mouse's brain.[22] They had available tools Lettvin couldn't have imagined. Where Lettvin had one crude electrode jammed into the optic nerve, the thick bundle of ganglion cell axons carrying their spikes to the rest of the brain, Baden and friends recorded directly from each one of hundreds of ganglion cells at the same time, and recorded over 11,000 in total. Where Lettvin showed his frogs a collection of randomly chosen objects and his "bug," a black dot moving across a metal dome by a magnet held by Lettvin himself, Baden and friends subjected every one of their ganglion cells to a barrage of computer-controlled light displays, each different element of the display designed to test one aspect of the possible changes in light—where it is, how fast it changes, what pattern it changes in, and what color it is.

By grouping their 11,000-plus set of neurons into those with similar responses to that barrage of input, Baden and company revealed at least thirty-two different types of ganglion cell. Some respond to the sudden onset or sudden offset of light. Some to different frequencies of change; some to different amplitudes of change. Some care about the direction of the light's movement, and some do not. Some respond in dim light, and some in bright. Some respond to the thing they care about with a brief burst of spikes; some respond with a continuous train of spikes. And whatever they respond to, each type of ganglion cell tiles the retina, so its very particular processing can be done on light coming from anywhere in the world that the eyes can see.

What are all these different types for? They each exist for one of two reasons. Either they are very selective about what they respond to in the world, and so exist to send messages that solve very specific problems. Or they are not selective, and each type of ganglion cell takes care of one aspect of the world that is very common.

A sterling example of a very selective type is the ON cell that only responds to light coming on quickly and moving in a specific direction, like left to right. While these ON responses are very useful for detecting when your mate is using a flashlight to send Morse code messages across a festival campsite when their phone has died ("G-e-t m-e a b-e-e-r"), unsurprisingly this is not why these types of cell evolved. One of the reasons the direction-selective ON type did evolve is to cope with your wobbly head. For example, if you want to keep looking at something while walking, your brain has to move your eyes down then up to correct for the up then down motion of your head, which is moving your eyes. How much your head is moving your eyes is worked out using the signals coming from these direction-selective ON cells:[23] as the eye moves up with the head, so the light coming into the retina from the object you're staring at will move down; so when those up-and-down selective ON cells farther down the retina start sending spikes, the brain knows how much the eye has moved and can then correct this eye movement by sending signals to the eye muscles to move the eye downward (and vice versa for when your head comes back down again).

Most ganglion cells types are not selective to narrow combinations of things needed to solve specific problems of coordinating the body. Rather, they each take care of a feature of the world that is common to everything we look at: small or large, fast or slow, edges and curves, color, and brightness. And the features of the world retinal neurons will care about depends on what animal they call home.[24] Different species are, well, different in some way: small or middling or large, hunter or hunted, active in the day or night or dawn and dusk, at home in the cold, tepid, warm, or hot, denizens of forest, grassland, desert, tundra, snow-field, mountain, river, shallow sea, or deep ocean. And each of those different ways of living demand different information reaches the brain via the eyes. It's no good having a retina full of neurons excellent at spotting green leaves of the forest canopy when you live in the deep ocean and eat plankton.

While we know the features ganglion cells like best from studying the retina of the mouse, we also know your retinas must be gathering more and different information. For example, we know you have a few types of ganglion cells that mice don't because your eye has three types of cones (which we call red, green, and blue), and mice have two, so your retina has ganglion cells that deal with information simply not available to the mouse's eye. But we also know that if we define types of ganglion cells by what genes they express, rather than the features of the world they care about, then you have fewer than mice: just twenty compared to the mouse's mighty forty (how these genetic types match up to the 30-plus feature-types found by Baden and friends is unknown).[25] Another big difference is that you have a fovea, an ultradense patch of cones, and mice don't. When you "look" at something in the world, that's you moving your head and eyes so that the photons fall on the cones in your fovea. This ultradense patch needs ultradense processing, which means both a dense clustering of ganglion cells there compared to the rest of the eye, and some types of neurons unique to the fovea. Compared to mice, your eyes send a few crucially different bits of information about the world to your brain.

All this means that as your eye falls on the cookie, the retina is splitting that cookie and its surroundings across tens of separate channels of information, each carrying different messages of cookie bits—the roundness of the cookie's edge, the brownness of the chocolate chips, the angle of the cookie box lid—to your cortex. And carrying messages of where the bits are in relation to each other; and messages of what direction they are in—for as you scan your head from left to right, and take in the cookie box, so the ganglion cells that respond to light moving from right to left are most excited (with me?—the light moves across your retina in the opposite direction to which your head is moving). That tumult of messages is shot down the ganglion cell axons, at least one million of them, all bundled together into the big white rope that is the optic nerve. To find out what happens to that message, we grab a spike as it shoots past, clinging on as it speeds along the axon to the distant shores of cortex.

CHAPTER 3

Legion

Our spike bursts into the first vision area of the cortex, V1, the first of the many areas dedicated to seeing that make up one-third of your entire cortex.[1] Its message—about one small pixel of crumbly chocolate temptation—needs passing on upward through all these areas, combining with all other messages carried by the millions of other spikes to create the perception of "cookie."

First we have to make landfall. Your cortex is a delicately layered cake, six layers in all, five layers crammed with juicy neurons, the first, top layer bereft of them. We're about to hit axon's end in the fourth layer of this V1 area. Above us the packed neurons of layers two and three; above them layer one, a smattering of rare small neurons, but mostly filled with axons going elsewhere and brain cells that aren't neurons, the glia cells, who are the scaffolding, the moppers-up, the below-stairs workforce. Below us, we can make out the larger, hulking neurons of layers five and six.

Their bodies may be packed into layers, but the bulk of the neurons are not. Around us it's a forest. A vast tree sprouts from the body of every neuron in sight. The tree's branches are thin but divide, ramify, contort. They fill a volume that dwarfs the neuron's body. This tree—the dendrite—is how a neuron collects its inputs from other neurons and funnels them all toward its body. Axon terminals from seemingly countless other neurons pepper the gnarled trees of neurons all about us.

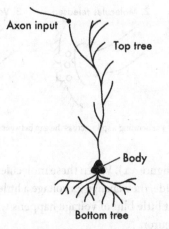

FIGURE 3.1. A pyramidal neurons of layer five of the cortex. So-called because its body is roughly pyramid shaped (in three dimensions).

The shape of the tree, and how many trees, tells us a lot about what that neuron is trying to do. Indeed, historically, it was often how we could tell neurons apart. Our trip from the retina is about to land us out in the compact, starburst tree of the first neuron in the cortex.[2] Below us, the poster-child neuron of the cortex, the pyramidal cell of layer five with its two types of dendritic trees—one sticking out of the top, a single long slender stem stretching up almost to the cortical surface, the rest sticking out below the body, fat and squat (figure 3.1). Above us, in layers two and three, a more modest pyramidal cell, its tree compact and surrounding the body, less attention-seeking than its big brother in layer five. Whatever shape and size,[3] all these dendrites are covered in inputs from other neurons.

But just as we plunge deep into layer four and hit axon's end, the spike's journey comes to a screaming halt. Between the spike and the next neuron is a gap, which it cannot cross. How can the spike's message be carried onward? How do we cross that gap and make a new spike in the next neuron to regenerate the message?

Our spike's arrival rips open bags of molecules stored at the end of the axon, forcing their contents to be dumped into the gap, and diffuse

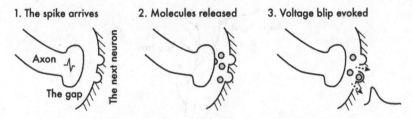

FIGURE 3.2. Sending a spike across the gap between neurons.

to the other side (figure 3.2). When these molecules lock into the neuron on the other side, they change its voltage a little. But only a little, a smidgen. And that little blip of voltage happens way out on a distant limb of the next neuron.

Exactly what effect each arriving spike has on the dendrites depends on what type of molecule is sent across the gap. A neuron has the same bags of molecules at each of its axon ends. And all neurons of the same type have the same bags too. But different types of neurons have different bags of molecules—and the kind of molecule determines whether the voltage will flicker up or flicker down in the target neuron.

Our cookie-pixel spike crashing into axon's end rips open bags of glutamate molecules. Bags split, the glutamate molecules tumble out of the terminal, diffusing across the micrometer-sized gap and bumping up against glutamate-shaped receptors on the other side. If by chance a glutamate molecule is oriented the right way around, it will lock snugly into the receptor—the whole process reminiscent of a two-year-old solving a puzzle by randomly mashing together the pieces and occasionally getting a nub sticking out of one piece to fit into a cutout of another. Random as it is, locking into the receptor triggers the opening of holes in the neuron's skin around it. The ions allowed through those holes create the blip of voltage in that bit of the target neuron's dendritic tree. And because it's a receptor that wants glutamate, the flow of ions creates a small increase in the voltage at our location on the target neuron. Excitation, we call it.

Nearby, farther down the tree, closer to the target neuron's body, we can see axon ends that have not come from the retina. Rather, they have been sent from small, rare neurons nearby. And these send across the gap a different molecule, GABA. When GABA locks into the GABA-shaped receptors on the same tree, the voltage flickers downward, decreasing. Unsurprisingly, we call this "inhibition."

When the blip of voltage goes up or down at a particular gap between axon end and target tree, that blip passes down the tree from the gap to the target neuron's body. There, decaying in size as it goes, this blip adds to the many flickers of voltage at the target neuron's body that contribute to making a spike: a blip *up* makes the target neuron more likely to create a new spike, a blip *down*, less likely.

The whole process seems a bit bonkers. Your brain went to all that effort to make a spike—a process that costs a lot of energy—to get around the fact that sending messages long distance can't be done just by dumping molecules or spreading flickers of voltage. And then it turns the spike back into dumped molecules, which cause flickers of voltage.

There are good reasons for doing this. For example, spreading voltage and chemistry are much cheaper in terms of energy—in tiny brains, everything is sent by spreading voltage and diffusing molecules, not spikes. But perhaps the key reason is flexibility. Transforming spikes back into chemistry then voltage gives the brain options for how to interpret the all-or-nothing spike.

This flexibility comes from the gaps having different strengths. Gaps of the same type, using the same bags of molecules, do not cause the same size blip of voltage. The blip is larger at some gaps than others, all other things being equal. This variation in blip size can arise by tweaking things either side of gap. For example, the target neuron across the gap can have more receptors that accept the molecules: the more receptors that get locked in, the more holes will open, and the bigger the voltage blip. And to the astute reader it should thus be obvious that we can also get a bigger voltage blip by simply dumping more molecules in the gap, to increase the number that are the correct way around by chance, and so increase the number of receptors they lock in to. Which all means the arriving spike can be transformed from its all-or-nothingness into a range of effects on the target neuron, from weak to strong.

But there are strict limits to how strong one gap can be. The whole shebang—the axon end, the gap, the receptors on the other side—is a mere few micrometers across. In that space there can only be so many receptors; the axon end can only store so much of the molecule. Those strict limits mean that one arriving spike is not enough to make a new spike.[4] For this is why we arrived on but one of millions of spikes streaming from the retina: we need a legion of spikes to make a new one.

THE MANY

The creation of a single new spike is the result of many other spikes arriving at a neuron, each little blip they cause adding up, accumulating, combining, until, finally, that neuron reaches its tipping point and spits out one spike. If you were a neuron in the cortex, spikes would seem legion. An endless barrage arriving, dumping their chemical load, and flickering your voltage up or down. This legion is essential, the many spikes to make one new spike, to carry the message forward.

How big is the legion? How many spikes exactly are needed to make a new spike?

We can get some ballpark answers by counting the number of inputs to a neuron. In the 1980s, Valentino Braitenberg and Almut Schuz painstakingly counted the number of inputs onto cortical neurons in a mouse.[5] They came up with a figure of about 7,500 inputs onto one cortical neuron. Clearly each of those inputs could not cause a spike by itself, otherwise the cortex would drown in spikes. But more than one and fewer than 7,500 spikes is still a pretty broad answer.

Thinking about the type of inputs lets us narrow the numbers down a bit. Remember, the inputs at some gaps cause the neuron's voltage to go down, not up. They inhibit the neuron, making it less likely to birth a spike. So we are really asking about how many of the inputs that excite the neuron we need to make a spike. Braitenberg and Schuz painstakingly counted those too—dedicated, admirable scientists, but ones who'd monologue at you for three hours on the best way to slice a mouse brain into wafer-thin bits and count its synapses with nary a pause to sip the steadily warming beer on the table leaving you trapped, beer-less, in an etiquette nightmare. They counted: about 90 percent of the inputs to a cortical neuron excite it; only about 10 percent inhibit it. We've brought our upper limit of the number of spikes down a bit to 6,750. Woohoo. I did say "a bit."

You'd think this would be an easy question—just tot up the number of arriving spikes needed to make the neuron's voltage reach the tipping point. But it's a hard question to pose to real neurons, because we have no feasible way of monitoring the thousands of inputs to a single neuron at the same time. Some have tried to get around this problem. In an elegant experiment, scientists in Michael Häusser's lab recorded from one cortical neuron while forcing one of its input neurons to fire a spike.[6] By doing this over and over again, they found that the single extra input spike increased the chances of the target neuron making its own spike by about 2 percent. Implying that if we wanted a cast-iron guarantee of making a spike, we'd need about 50 extra input spikes. Extra—on top of the inputs the neuron was getting anyway. We've raised our lower bar to about 50 spikes, and the top is still 6,750. Can we do any better?

Totting up input spikes is an easier question if we instead pose it to a pretend neuron. We have many flavors of pretend neurons we can write down in math terms and simulate on computers. Hodgkin and Huxley wrote down one of the foundational models to prove that the opening and closing of holes in a neuron's skin would indeed create the voltage spike in an axon (and simulated that model—four complex and coupled equations—using a hand-cranked calculator and a pencil). Their 1963 Nobel Prize was as much for bloody-minded perseverance as genius. So we can take one of our pretend neurons, send it pretend spikes as inputs to its pretend gaps, and ask: how many inputs do you need to fire?

And they tell us: it depends. Roughly? 100 to 200. Roughly, if we take a complex pretend cortical neuron, one with full pretend trees, and full pretend receptors for the pretend transmitted molecules, and make all the spikes turn up at the about same time, then about 180 arriving spikes are needed to guarantee one new spike.[7] But that's if we ignore a lot. Like exactly when the spikes arrive relative to each other—spread out in time, or all bunched up. And that there will be spikes arriving all the time, so when we should start counting is unclear. And spikes arriving at the gaps that inhibit the neuron. And what the relative strengths of all these gaps are, for the stronger they are, the fewer will be needed. And whether the voltage blip produced at the gap lasts a short or long time. All for just one particular type of neuron in the cortex, the pyramidal neuron.

Because in truth "how many spikes?" is a deep, hard question whose answer depends on myriad factors. And these myriad factors tell us a lot about how the brain uses spikes to make things happen. Three stand out: the balance of excitation and inhibition arriving at a neuron, the synchrony of the inputs, and where they land on the tree itself.

THE GOLDILOCKS ZONE

The legion of inputs is dangerous. A few hundred spikes is enough to birth a new one, but these are spread over thousands of input lines. Worse, across these inputs, excitation outnumbers inhibition by at least five to one. Even just a few extra spikes across these thousands of inputs leads to a runaway loop—of spikes triggering spikes triggering spikes—that would crash the brain. Epilepsy is one such crash: massive waves of spikes across the cortex, so many spikes that every neuron on the receiving end of a wave immediately reaches its tipping point—each birthing a spike at the same time, and making the next wave.

But such crashes are rare. They're rare because the brain is kept in its Goldilocks zone—not too active and not too quiet, but just right.[8] And it stays in this zone by keeping the perfect balance between excitation and inhibition.

We uncovered the brain's balancing act from a simple puzzle about the spaces between spikes. In 1992, William Softky and Christof Koch reported something was amiss in the spikes being sent by neurons in the first visual area of cortex,[9] the exact same kind of neuron we're making a spike in right now. Wading through hundreds of recordings of single neurons, they noticed the spikes coming out of each neuron were created at remarkably irregular intervals. A short interval between spikes could be followed by another short interval, or a medium interval, or sometimes a long interval. Or any combination thereof. Indeed, for some neurons the ordering of the intervals was almost perfectly random, so that if you took the intervals between their spikes and shuffled them into a different order, you couldn't tell which was the original set of spikes and which was the shuffled set of spikes.[10]

As theorists, they instantly realized something was amiss. Our best models for how neurons make spikes don't have randomly different intervals between those spikes. No matter how irregularly spaced the spikes these models receive, the spikes they make are evenly spaced, the intervals between them far more regular than Softky and Koch saw in the cortex. To grasp why, think about the total number of spikes arriving at a neuron. Even though each of the individual inputs has highly irregular spikes, there are thousands of such inputs. So when we sum over them to get the total number of spikes arriving, then we find the sum is pretty constant in time. So if a model neuron needs, say, 175 input spikes to make one new spike, then a total of 175 input spikes turns up regular as clockwork, thus making a new spike regular as clockwork (figure 3.3).

According to our best models, spikes coming in irregularly would be made into an output of regular, well-behaved, evenly spaced spikes. But this creates a paradox: if neurons make regular-spaced spikes, where then do the random, irregularly spaced spikes of cortex come from in the first place?

Theorists love paradoxes. Paradoxes in science show us where there is a gap in our understanding, and hold out the promise that solving the paradox will create a new view of how the world works. Fittingly, the irregular-spike paradox invoked a pile-on of theorists, and a raft of proposals for what could be making irregular spikes.[11]

From this churn, the balanced-input theory emerged dominant. Once the paradox was posed, Michael Shadlen and Bill Newsome quickly pointed out that irregular spikes would be guaranteed if the total amount of excitation and inhibition into the neuron varied randomly but were about the same on average.[12] That is, if excitation and inhibition were balanced: some neurons delivering irregular spikes that excite their target neuron, and some delivering irregular spikes that inhibit the

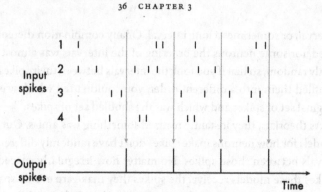

FIGURE 3.3. How randomly spaced input spikes make regularly spaced output spikes. Imagine we were sitting on a neuron receiving inputs from four other neurons. Each of their trains of spikes is shown above: each tick is a spike, each row of ticks the spikes from that neuron sent to the one we are sitting on. Each row is quite random: the spaces between spikes are long and short, with no clear order. Now imagine that our neuron needed seven spikes in total to make it spike. We count up the spikes arriving from the four neurons and mark an output spike every time we get seven in total (the gray lines). The resulting output spike train is regular, because a total of seven spikes across the four individually random inputs is a regular occurrence.

same target neuron. Which cancel out, but not exactly. Because the flickers up and down of voltage will be random, sometimes the flickers up will be big enough to reach the tipping point. And, randomly, a spike is born, giving random, irregular gaps between spikes.

Great theory. But we've just learned that excitatory inputs vastly outnumber inhibitory inputs. So for this theory to work, it makes some really strong predictions. Canceling out excitation means that either inhibitory neurons send far more spikes (so the total number of spikes to exciting and to inhibiting inputs is balanced); or that the inhibitory inputs should have a greater effect per spike (so the total amount of voltage is balanced); or some combination of both. There is now evidence for both.[13] Local neurons in cortex that make the inhibitory inputs onto our neuron fire two- to threefold more spikes. And the gaps where these spikes arrive can be four- to fivefold stronger than those of excitatory inputs. So the balanced-input theory explains why your cortex doesn't crash: the cortex is set up so that the total amount of inhibition is just right to cancel out the total amount of excitation.

But so far that's just the input to one neuron. To know that the brain is indeed in balance, we need to know if a network of neurons can be kept in balance. That if we wire a bunch of pretend neurons together, many excitatory some inhibitory, then they can all make irregular spikes and so be each other's irregular inputs. It is not obvious this will work. For example, imagine that a neuron's output spikes are irregular but just a little more regular than the input. If each neuron's output is always a little more regular, then the whole network will eventually collapse into a state where all spikes are sent regular as clockwork. The victory for theory here was showing not just that such balanced networks can exist,[14] but that they can bring themselves into existence, can self-organize.[15]

The math is fierce, but the idea is simple. We've got our pretend neurons, most excitatory and the rest inhibitory, and we've randomly wired them together. Then all we need to do is guarantee that the input to each neuron is more than the neuron needs to make a spike. For this then creates a web of negative feedback loops, of neurons trying to make a spike, but being held back. It works like this. Say some of the excitatory neurons send a lot of spikes. This will drive inhibitory neurons to make spikes—that are fed back to those excitatory neurons and turn down their spikes. But they can't turn all the excitatory neurons down too much, because then the inhibitory neurons themselves will lose their input and stop firing. But they have to be firing, because the excitatory neurons are firing. This paradox implies there is a self-consistent state for the whole network, where the inhibitory and excitatory neurons are sending just the right amount of spikes each, so the push-pull of the excitation and inhibition in the network is balanced. And we all know by now what balanced excitatory and inhibitory inputs means: irregular spikes. Better still, this theory showed balanced networks are robust old things: you don't need fine-tuning of the exact strengths of inhibitory and excitatory inputs; nor do you need to fiddle with the details of how the neurons make spikes. Just make the total input to each neuron a lot, add a dash of feedback, and voilà: balance.

Experimental data then poured in to test these ideas. Recordings inside neurons from across different bits of the cortex, from the bit that deals with a rat's whiskers,[16] to the bit that deals with seeing in ferrets,[17] to the bit that deals with hearing,[18] all showed the same thing: that the total amount of excitation and inhibition coming into a cortical neuron was roughly in balance.

These difficult, exquisite experiments threw up an unexpected twist, kicking the ball back into the theorists' court. The theories were about balance in general; that, on average, the total amount of inhibition and excitation across a circuit of neurons cancels out. But the data show that the balance seems to be maintained at the input to each and every neuron. Not just maintained, but absurdly precise: that as the amount of excitation shrinks or grows, so the amount of inhibition tracks it exactly.[19]

The tale of the irregular-spike paradox is a lovely example of science in full swing, a dialogue between theory and experiment, a burst of creative theories brought forth by a clearly posed problem. A dialogue that revealed the brain's Goldilocks zone. So we know that our spike, arriving from the retina and making its small upward blip of voltage, joins hundreds of other spikes arriving on the same tree, together evoking a torrent of blips up and blips down, kept in balance to make a new spike.

NEURAL ORCHESTRA

If you really want to make a neuron fire a spike, nothing is more effective than having its excitatory inputs all turn up at once. The more synchronized those inputs, the faster the voltage blips will accumulate, and the fewer you need to make a new spike. If you sat down to design a foolproof way for spikes to send important messages across the brain, you'd stick synchrony into your blueprint first. Synchronize the spikes coming into a neuron, and their message will be carried onward in the newborn spike.

If evolution followed the same blueprint, the brain should be a neural orchestra.[20] There should be choristers, neurons spiking together in harmony, carrying their message together. And perhaps soloists, neurons spiking in blissful isolation, elaborating on the central theme.

Wonderfully, there are. If we record lots of neurons at the same time, we can ask of each neuron its role in the orchestra. The joint lab of Matteo Carandini and Kenneth Harris at University College London, with Michael Okun taking the lead, came up with a disarmingly simple way to assign these roles.[21] They simply asked of each neuron: how much does your firing look like the average firing of the population you're sitting in? They found a continuum: choristers at one end, slavishly copying the population's rise and fall in activity; soloists at the other, following their own path, playing from the heart.

The orchestra metaphor implies harmony, implies that the soloists are in the same key as the chorus. But no: in the cortex, the soloists are unrelated to the chorus. Think Miles Davis cutting loose. And the choruses are not related to another. Indeed, at times the cortex is closer to Ligeti than Handel, to the awesome noise of *Kyrie* erupting at our first sight of the monolith in *2001*, piles upon piles of voices, choruses in separate harmonies, soloists drifting in and out untethered.

Choristers can come in different flavors. Some choruses of neurons are created by them responding to the same thing. When your eye falls on that squidgy round bit of pale pear sticking out the top of the cookie, you want that message to get from your eye to your brain intact. Neurons from the retina that care about that pattern of contrasting light in that location of the world will be sending spikes together. A chorus of spikes screaming "mmmm pear!" in C sharp.

Our spike is then part of a transient chorus, of spikes shot en masse from the retina, from the same type of ganglion cell, in the same location in the retina, arriving at the same first neuron in the cortex. And as you might imagine, if that chorus turns up at our single neuron in cortex, that neuron's spike will reflect that message.

Which is why the neuron our spike landed on is called a simple cell. For they like simple things.[22] They like light and dark patches of the world to fall next to each other at a particular angle. Or "edges" as we call them—some will like the edge created by the light brown of the cookie against the dark brown of the lid; some the dark brown of the lid against the diffuse light of the office; some the diffuse light of the office against Graham's hideous black and purple striped shirt that he insists on wearing every single Tuesday apparently to remind everyone the weekend is a long, work-drenched way off. A simple cell sends its spikes when it sees what it likes. And what it likes is defined by the hundreds of inputs arriving from the retina. For simple cells to respond to one particular thing means that their inputs must be mostly about this thing, must be a chorus.

There are two reasons the chorus we've arrived in is important. It's important that the chorus harmonizes information. Remember that the varied types of ganglion cells split into those sending spikes to patches of dark (OFF cells) and those sending spikes to patches of light (ON cells). The simple cell responds to a particular combination of dark and light patches, so its chorus must contain the contributions from OFF cells in one location and ON cells in another location next door.

It's also important that the chorus arrives roughly together, for inputs from the eye are vastly outnumbered by those from other neurons in cortex. There are far fewer inputs feeding direct information from the eye than there are inputs feeding back spikes from other neurons in cortex. So for the simple cell to respond when information comes from the eye, those spikes from the eye need to come at about the same time, to drive the voltage up to the tipping point.

Other types of chorus we'll see more of as we hurtle into the deeper reaches of the brain. One—further torturing our musical metaphors—is an ensemble, a collective of neurons who always send spikes together. And not just because something out in the world made them do it. They send spikes together even when recorded in a slice of brain stuck in a dish. Another flavor of chorus sends spikes en masse occasionally, unreliably—like a primary school choir with poor concentration skills, different members piping up at any one time.

All types of chorus dramatically increase the chance of producing a new spike in a neuron on the receiving end.[23] Which is why a chorus was also proposed as a solution to the irregular spike paradox. In this solution, irregular gaps between the spikes sent by a neuron came because its inputs were irregular but synchronized, so each "ahhhh" of this input chorus, whenever it randomly turned up, would create an equally randomly spaced set of output spikes.[24] Indeed there is some truth to this; it turns out that self-balancing networks automatically create some synchrony to the inputs of a neuron.[25] But, just like balance itself, the synchrony has to be just right: too little, and no effect; too much, and we crash the brain.

Balance and synchrony also intersect to create precise spikes. Neurons in cortex reliably see a few milliseconds delay between an increase of their excitatory input and the matching, balancing increase of their inhibitory input.[26] A delay that seems tailor-made for allowing a chorus of excitatory inputs, like the barrage we've arrived in from the retina, to make a single, precise spike, before being shut down by inhibition.

But the precise effect of the choristers is not guaranteed. For it depends where on the neuron their axons join.

"TO SUM UP"

Balance and synchrony are properties of the inputs to a neuron. But the neuron's own dendritic tree also plays a key role in the birth of its own spike. Where each arriving spike lands on the neuron can control precisely how big a voltage blip it will make and so determine exactly how many spikes we need to make a new one.

Befitting the cliché of location, location, location, there are three ways the tree influences how many spikes we need. The first is how far away from the neuron's body the spike lands; the second is how bunched together on the tree are inputs from a chorus of neurons; the third is what lies ahead on the path between that input and the body.

Spikes leaping gaps to land on the tree of a neuron can be far out from body, up to a millimeter away. But a voltage blip all the way out there decays rapidly as it travels down the tree to the body, decaying to almost nothing by its destination, contributing little if anything to the neuron reaching its tipping point. Blips created by spikes arriving at gaps close to the body decay little, so can have a big effect on moving toward—or away from—the tipping point. Bit of a head-scratcher from evolution there; it seems a tad pointless to have inputs way out on tips of the tree, yet there they are.

Bunched-together inputs come to the rescue here. Big neurons in the cortex, like the layer five pyramidal neurons just below us, rudely jutting their trees past our simple neuron in layer four right up to the ceiling of cortex in layer one, those ignoramuses, have a trick up their sleeve. They add up wrong.[27]

One or two spikes arriving close together on a single branch of these neurons make a standard, small blip of voltage. But three or more spikes arriving together on the same branch make a massive voltage blip. A blip bigger than simply adding up the separate blips made by each spike arriving alone.

This supralinear sum is a sudden jump of voltage in the branch where the inputs land. Enough inputs arriving together opens up new holes in the neuron's skin, allowing extra ions to flow into the neuron, driving up the voltage in that bit of branch. And if you're thinking that sounds like a spike, you'd not be far off. While not a pretty, peaky thing, this sudden jump in voltage in the tree has the same job: get information from the far reaches of the tree down to the neuron's body intact. So if a bunch of inputs turn up at same time, they evoke this superblip that rushes down the tree and makes a major contribution to the neuron reaching its tipping point.

We can plug this superblip into our pretend neurons to ask exactly how much it contributes to making a spike. To do that, we make each set of inputs carrying synchronous spikes bunch up together on one distant branch of the tree. Each volley from a chorus will then tend to make a superblip, in one branch of the tree. With this bunching, even with the inputs far out on the tips, we need as little as a third of the spikes to make a new one, compared to the spikes arriving spread out over the tree.[28] Location and synchrony working together can dramatically increase the chances of making a new spike.

Inhibition is the key player in the third part of the location trilogy. For ahead be dragons. On the path between where our spike landed on the tree and the neuron's body lie many other inputs. Many of these are other exciting inputs, invoking more upward blips of voltage. Friends to help us in our way. Some though are inputs from neurons swilling GABA, each of their spikes making downward blips of voltage. Big downwards blips. If those GABA inputs between us and the body are evoked just before we pass them, our blip would be canceled out, annihilated.[29] We'd never reach the neuron's body.

Worse, we might never see it coming. Inhibition has a ninja mode: silent, unseen, deadly. You see, the size of a voltage blip made by an input also depends on the voltage already in the branch where it lands. This is particularly true for GABA-driven inputs, for there is a level of voltage, the "reversal potential," well within a neuron's usual range, at which the GABA input creates no voltage blip at all. For at the "reversal potential" no ions are flowing through the opened holes in the neuron's skin. But the GABA is still there, locked into the receptor, holes open, ready to make ions flow. So when an excitatory, upward blip of voltage tries to pass by, it drives the local voltage away from this reversal potential—and the ions start to flow through the already open holes. The GABA input, invisible to voltage, drains the passing excitatory blip as the ions ebb away, and shunts it to nothing. Come to think of it, perhaps ninja is the wrong metaphor—vampire is more apt.

These location, location, location dependencies of a single neuron have deep implications for artificial intelligence (AI). AI-brand neural networks are all constructed from the same kind of pretend neuron, a simple thing that just adds up its inputs from other pretend neurons. And once added up, an AI-brand pretend neuron checks if they sum greater than zero, and if so then sends that sum on to all its targets (or else sends zero). The deepest of deep networks are all constructed from millions of these elementary things. But I've just spent five thousand and more words telling you that a single neuron in your cortex does not just add up its inputs. How a neuron deals with its input depends on a plethora of things that all interact: the balance and synchrony of its inputs, where they land and how they bunch, whether they sum up wrong, what the voltage is when the inputs land, and what is on the path between the input and the neuron's body. Current AI networks have barely tapped the surface of what brains do.

Indeed, detailed pretend models of neurons have shown us that a single neuron can itself be a two-layer neural network.[30] If each branch of the tree has the ability to do those supralinear sums, each branch acts as single AI-brand pretend unit. So their outputs (layer one of the network) converge at the neuron's body (layer two of the network). And with this comes the ability for single neurons to compute alone many of the functions of logic.[31] So each individual neuron is close, in the set of computable functions, to the laptop sitting in front of you, cursor flashing on the empty first page of the monthly reports, its tilted overbright screen thankfully not obscuring the flat, squarish box holding the just-noticed cookie conundrum. Each neuron in cortex is, approximately, a computer.[32]

It turns out that making a new spike is a fiendishly convoluted business. Our spike from the eye jumped the gap, transmuting from an all-or-nothing wave of voltage to chemistry and back to a small blip of voltage. With us, many more spikes arrive from the eye, whose blips drive the neuron toward its tipping point. Our spike is one tiny part of a brash chorus singing the praises of cookie-in-a-box-on-the-desk. Outside our chorus, a legion of other spikes arrive, from local neurons spewing GABA, whose voltage blips drive down the neuron, away from its tipping point. Sweeping down the tree toward the neuron's body, the convergence of up and down blips is in balance to keep the brain just right. Yet as they sweep down the tree, we've watched some of our fellow travelers fade and die, others killed suddenly by GABA ninjas. Through the maelstrom, suddenly, randomly, a series of upward blips arrive together, driving the voltage to the tipping point, and a new spike is born from the first neuron in the cortex. To go—where?

CHAPTER 4

Split Personality

DIVIDE AND CONQUER

Spikes go where axons go. Each axon erupts as a gossamer-thin cable from its neuron's body, opening the line of communication from one neuron to the next. When we think of connecting two neurons, the axon may seem a single strand of wire, tying one neuron to another, a private line of communication, two tin cans on a string. But the axon is not one dedicated line, carrying the spike's message in private from one neuron to an intimate partner.

The axon is a tortuous structure, splitting again and again, branching furiously, winding and twisting and tumbling. Ahead of us, the simple cell's axon twists and turns, dividing in two over a hundred times. Branches upon branches, but not evenly spaced: some are next to the body, some in the layers above and below us, some far distant. And hardly unique to our simple cell—most neurons in the cortex sprout torturous rivers of axon, covering a volume that dwarfs the parent neuron and its tree. Indeed, towering over our simple cell, we can see above us the massively branched axons of the pyramidal neurons in layer three, dividing hundreds of times, one long branch cascading down past us to then itself split again and again below us in the fifth layer. We barely have time to take this in as our spike hurtles into the first fork in the river of axon before us.

At every split our spike is copied, cloned, sent down every new branch to carry its message onward. Down hundreds of branches. So from one spike shot from a neuron's body are created hundreds of clones. Clones that trigger the release of molecules at gaps all along the branches. For gaps between our axon and the tree of another neuron are strewn along its entire length. In some places we flash past these gaps every five micrometers, packed as tightly as their molecular machinery will allow, each one primed to dump its molecular load and trigger the blip of voltage on the other side. Other long stretches of axon flash by with no branching and no connections to another neuron, often weirdly straight, these Roman roads of axon dedicated to getting the spike to another place, to another wild burst of dividing and twisting.

By axon's end, our spike has made contact with thousands of other neurons. On the other side of most gaps is a different neuron each time. So most of the roughly 7,000 excitatory inputs onto one cortical neuron are made by different axons, arising from different neurons. Meaning that each excitatory neuron in cortex connects via its axon to roughly 7,000 different targets, a wide sampling made feasible precisely by the twisting, turning, tumbling course of the axon, as it kinks and spreads to move away after each contact, to hit a new target.

By twisting, branching, and cloning, a neuron's axon broadcasts a single spike to thousands of listeners across the cortex. Who those listeners are and where they sit in the brain tell us a lot about that neuron's job. To fully understand what message our spike is carrying we need to know two things of the neuron that created it. We need to know what the neuron responds to, what created that spike in the first place—this is the legion. And we need to know where it sends that spike, to whom it is being transmitted. For an axon can feed the spike it carries everywhere in the cortex. Nearby, to recruit more neuron kin to amplify and clarify the message. Far, to carry its message across cortex to other areas, there to be combined with other message-laden spikes, creating ever-more-complex representations and computations. And farther, across the hemispheres, to keep the brain in sync.

NEAR

The first target of our spike is other simple cells. Simple cells surround us here, where we've landed in the middle layer, layer four, of the first vision area of the cortex. The axon of our simple cell snakes out, branching repeatedly around its own body, each branch brushing past the tree of another simple cell. But as we follow each cloned spike down those branches and across the gaps, we find many simple cells on the other side are strikingly similar to the one we just left. They like the same thing.

Belying their name, simple cells are an eclectic bunch. For one thing, they keep the orderly map of the visual world that came from the retina, so nearby simple cells respond to similar positions in the world. For another, tens of channels of information from the retina have slammed into the simple cells that surround us. Thirty-plus channels for every location in visual space, for the middle, the left, the right, up and down, everywhere. So collections of simple cells bunched together are interested in different things about the same location in the world: some in edges at 90 degrees, some at 120 degrees, some at 41.3 degrees; some in edges created by a patch of light above and dark below, some reversed. And all combinations in between.

So if our simple cell just stuck out its axon at random, then our spike should be equally likely to find its way to any of this eclectic bunch. But it isn't. We know this because researchers in Thomas Mrsic-Flogel's lab at University College London have been tracking the local destinations of single spikes in this first bit of visual cortex, in exceedingly clever experiments.[1] They recorded hundreds of neuron at the same time from the visual cortex of mice sat down in a mouse cinema, watching slide shows and movies, all so they could work out from the recorded activity where and what each neuron liked about the stuff happening in the visual world.

Now knowing what each neuron liked—the neuron's tuning—the scientists in Mrsic-Flogel's lab needed to work out the connections between them. They switched to a delicate, difficult technique for recording every flicker of voltage from a handful of neurons, four at most, at the same time. Delicately positioning an electrode directly onto the skin of each neuron's body, a microscopic dot ten times narrower than the width of a human hair, they sought solid evidence for a direct connection from one of those neurons to another by evoking a spike from one neuron and checking for a subsequent blip of voltage in the others. No blip, no connection, but reliable blips in neuron Bertha after evoking

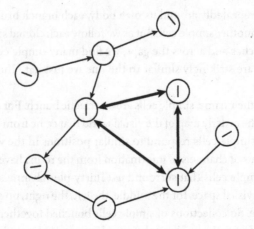

FIGURE 4.1. How neurons in V1 connect nearby. Each circle is a neuron in V1; each arrow between them is a connection from one neuron to another. Within each circle there is a line showing the angle and position of the edge that the neuron most likes. Neurons that like similar edges have stronger connections (thicker arrows) and more connections to each other. Connections between neurons that like dissimilar edges are weaker and fewer.

a spike in neuron Aleph are pretty damn convincing evidence that neuron Aleph connects directly to neuron Bertha. Having found a connection, Mrsic-Flogel and team could then go back to their mouse cinema data and ask: what did these neurons care about in the visual world?

Complex work, with a simple conclusion: the probability of finding a connection between two neurons with similar tuning is vastly greater than chance. Neurons that like very similar things at very similar positions in the visual world also like to connect together. And not just connect. Connect strongly. For the strength of the connection between two similarly tuned neurons, measured by the size of the voltage blip, is also vastly greater than chance would predict[2] (figure 4.1).

Our spike came from a simple cell that responds to edges at an angle of 30 degrees in the top right of your view of the world, where the tip of the lid of the cookie tray happens to be, conveyed by many channels of retinal outpourings. The local branches of the axon send the clones of our spike straight to simple cells with the same tastes, the same love

of 30-degree edges in the top-right corner of the world. And those neurons will do the same in return, sending one of their cloned spikes back
to the neuron we just left.

The work of those in Mrsic-Flogel's lab tells us how we've ended up
at neurons that like similar things in the world, but it doesn't tell us why.
Why is simple. Remember, it takes a legion of spikes to make one more
spike. So if we want the rest of the brain to know about 30-degree, top-
right edges, then it makes sense to recruit as many spikes as we can with
the same message, to cajole the neighboring neurons into sending their
spikes to join us in the forthcoming journey across the cortex.[3]

There are not just simple cells around us in this fourth layer of cortex.
There are many other neurons around us that like more complex combinations of information coming from the eyes; cunningly, they are
called complex cells. Complex cells send spikes when they see what they
like, and what they like are combinations of patches of light and dark,
with each pair of light-dark patches at a specific angle.

Sound familiar? Indeed, the simplest idea for how vision works is that
complex cells are made by combining inputs from simple cells.[4] That
each simple cell is a feature detector for exactly one type of edge, and a
complex cell combines the outputs of a few feature detectors into a
combination of edges. Our spike's clones thus also make their way to
the complex cells around us, carrying a message about one of the simple
features in the world the complex cell will add up.

This neat picture is a useful guide, but not quite accurate; a neuron
being simple or complex lies on a continuum.[5] Some neurons are plain
and simple, responding to a single edge at a single angle; some are perfectly complex, adding up the outputs of plain and simple cells. And
many are in between, being complex-like. Regardless, everything just
said above is true for any complex-like cell too; they are more likely to
link to other neurons with similar tunings for similar positions in the
visual world.

And while neurons in the first visual cortex are more likely to connect
to nearby neurons with similar tuning, they don't only connect there.
After all, compared to the total number of neurons immediately around
us in this fourth layer of visual cortex, the number of neurons that like

roughly the same kind of simple or complex edges in the same position in the world are very few. So most of the nearby neurons we run into on our cloned spikes, those on the other side of the gaps, will be not quite like the neuron we just left. Indeed, from the perspective of one of these neurons—including the one we just left—only a handful of its inputs are from neurons with similar tastes. As we've just seen, those inputs are vastly more numerous and stronger than expected, so the blips they create play a key role in driving the neuron's flickering voltage to its tipping point for a spike.

But still there are many others. And one of the marvels of modern neuroscience is that we can see these individual inputs in action, for we can video the effect of a single voltage blip in a tiny stretch of dendrite (to be precise, we attach a fluorescent molecule to calcium in the cell, and video the change in fluorescence: a single voltage blip will cause a change in the amount of calcium in the same bit of dendrite, so an increase in fluorescence means the input has just been activated—a spike arrived, dumped molecules, and induced a voltage blip). We can even do this videoing in a spine, a tiny little sticky-out stub of dendrite that has just enough space for a single gap between the axon on one side and the dendrite on the other, so we can be absolutely sure we're looking at a single, solitary contact between one neuron and another. And as the blip is made by a spike in response to something in the visual world, so we can ask what that single input cares about in the world.

When Sonja Hofer and her team at the University of Basel videoed the single inputs to neurons in a mouse's visual cortex, in another mouse cinema, they saw each tree was festooned with a menagerie of inputs from other, dissimilar neurons.[6] Most strikingly, a handful of inputs came from neurons that were not remotely looking at the same part of visual space. If they recorded the inputs to a neuron that itself liked things in the bottom left corner of the visual world, then a few of its inputs lit up when something was passing through the center of the world, a few others when something was passing through the top of world, and more besides.

So clones of our spike jump gaps onto neurons that are looking at the center of the box lid, at bits of the wobbly, upside-down scrawl of

"Cookies"; other clones jump gaps onto neurons looking at the dull expanse of the fake rosewood laminate desktop stretching beneath the box; yet others onto neurons staring intently at the exact point in space where the biggest chunk of pear touches a wedge of solid chocolate. And none of these inputs is enough to make the neuron on the other side create a spike, but they can change exactly when that neuron makes a spike. They provide context, allowing information in one bit of your view of the world to inform another bit.

There is one more type of neuron nearby, a type of neuron that sits in the very first bit of visual cortex but does not particularly care for the view. Everything we've landed on so far is an excitatory neuron, a stellate cell with its starburst tree or a pyramidal cell with its above-and-below tree. But as clones of our spike travel along some branches of the axon, they leap gaps onto rare GABA-toting neurons, neurons that get nothing from the retina, and only branch their axons inside the region of cortex their bodies sit in. Hence we call them interneurons. They are the wellsprings of the deadly inhibition we ran into when racing down the tree of our first neuron. They are there to sort the wheat from the chaff.

These interneurons get their inputs from the excitatory neurons all around them and send their spikes back to that same collection of excitatory neurons and more. Our cloned spike jumping gaps onto these interneurons is then attempting to deliberately increase the inhibition of other excitatory neurons. To suppress them. Indeed, very recent evidence[7] suggests that a single spike from a single excitatory neuron can weakly but detectably lessen the probability of a spike in neurons all around it, up to 500 micrometers away, almost certainly by harnessing the power of the GABA-toting interneurons.[8] Yet at the same time, as we've seen, that single neuron will weakly but detectably enhance the probability of a spike from those few like-minded neurons with very similar responses to itself.

Why suppress so many other neurons and enhance just a few like-minded ones? There's a simple theoretical answer to that: send only what needs to be sent. If many neurons with almost-but-not-quite the same tuning for the visual world send spikes across the cortex, the receiving neurons will be getting a lot of information that is both redundant and ambiguous. With our spike meaning an edge in the top right of the world at an angle of 30 degrees, sending spikes from other neurons saying there's an angle of 28 or 32 degrees in about the same location both wastes spikes and makes it unclear what exactly is in the world (leading to a fuzzy box lid). By harnessing the interneurons to suppress other neurons with kind-of-similar tunings, our spike is trying to stop these redundant and ambiguous messages, to conserve energy and create clarity.

It's time to leave this maelstrom of recurrence behind, of crisscrossing axon branches linking the cortical neurons of the landing zone. Clinging to one cloned spike speeding along a so-far-unremarkable branch of axon, we suddenly hit a sharp upward turn and launch into the layers of cortex above us. Here the axon branches and branches again, along each branch cloned spikes racing to land on the pyramidal cells across layers two and three, the neurons that start the long journey across the cortex.

FAR: HIGHWAY WHAT AND HIGHWAY DO

We've followed many cloned spikes to their destinations, watching them trip the circuit on gaps as they speed past on their branches, showering the trees of neurons on the other side with molecules that lock into receptors and trigger voltage blips now receding into the distance down the trees toward the target neuron's body. But now we're moving onward and upward into layers two and three, it's time for us to make the leap ourselves. We jump a gap onto a pyramidal cell in layer two, feel that familiar flicker of voltage far out in its tree, and slide down with it toward the neuron's body, there to join a cascade of blips pushing the neuron to its tipping point, sparking to life a new spike for us to tag along with, tracking its clones through the hundreds of branches of this pyramidal cell's axon.

On our route forward we complete the circuit between the layers of the cortex. From here, we could follow cloned spikes traversing axon branches in layers two and three to hit the trees of pyramidal cells from layer five, their long slender trunk diverging into a canopy around us. Or we can follow a cloned spike plunging down the axon, back through layer four, to hit the trees of pyramidal cells in layer six, at the bottom of the cortical stack.

In tracing this feed-forward circuit, from layer four up to layers two and three, and back down to layers five and six, we've run into all three types of pyramidal neurons in cortex.[9] All three types use glutamate as their molecules, so all excite neurons they connect to; all connect to neurons of the same type within their own layer, but they are separated by where they send the long branch of their axon. In layer five we watch some of the cloned spikes run into pyramidal tract neurons that send a long branch of their axon all the way through the brain, down to the brain stem, and some onward to the spinal cord. In layer six other cloned spikes jump gaps onto cortico-thalamic neurons, their long branch dropping down out of the cortex into the tiny nugget of midbrain called the thalamus, whose own neurons send their axons all over the cortex, creating complex feedback loops. And in all layers (except layer one) are the cortico-cortical neurons, whose long branch of axon connects regions within the cortex, carrying spikes far across the cortex on this left side of the brain, and on to the cortex on the other side of the brain. Neurons just like the one we've been dispatched from.

For now, we ignore the clones that jump ship to the neurons projecting beyond the cortex—we'll catch up with them later (much later—in chapter 8). We plunge down the long branch of the axon from the pyramidal neuron in layer two; down past layer four; past layer five; past layer six; and into the white matter, screeching around a neck-jolting ninety-degree turn to join the superhighway of axons that crisscross the cortex.

For seeing, decades of work has unraveled two of those highways in extraordinary detail: Highway What, and Highway Do[10] (figure 4.2). Sent down one highway of axons, our spike will jump from area to area to help create the message of "what": spikes carrying messages of curves, edges, browns, whites, and more are brought together to reveal the single remaining ginger, pear, and chocolate cookie, in its tray, the lid pointing away from you, on the edge of the desk, tantalizingly within

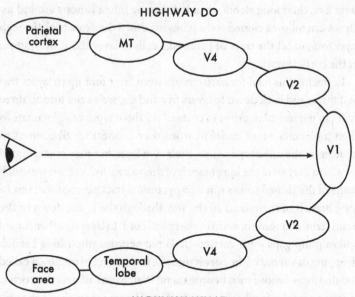

HIGHWAY DO

Parietal cortex — MT — V4 — V2 — V1

Face area — Temporal lobe — V4 — V2

HIGHWAY WHAT

FIGURE 4.2. The visual Highways of the cortex. Information from the eyes arrives in V1, then travels onward through the two Highways of axons that join the visual areas of the cortex. Every jump between areas is a plunge down an axon into the white matter, to reemerge in the next area along.

reach. Down another highway, the areas our spike leaps through will create the messages of what you need to know to "do" something: spikes carrying messages of the distance, size, and movement of the edges and curves around you to reveal that you could, in theory, move your arm to reach the stationary cookie and as you do so spread your fingers just wide enough to pick it up.

Highway What

We leave V1 behind. We exit the white matter along an axon ascending within the second area of cortex that deals with vision, the drolly named V2. Quick warning: the names don't get any better. Ahead of us lie a litany of cortical regions best known by their uniquely identifying jumble of letters and numbers, as though prisoners of a cerebral jail: "Will area 7a report to the warden's office immediately."

In V2 we follow our spike across gaps to pyramidal neurons that like conjoined edges, edges that are next to each other, touching, in a particular bit of visual space. Neurons that will create a spike to the long stretch of the upward-sweeping line of the side of the box lid, or other neurons that will respond to the angle made by the edges where the lid and the tray meet.

Again, what a neuron in V2 likes is defined by what its legion of inputs tells it about. Our spike is part of the legion arriving from V1, from simple and complex-like cells looking at adjacent bits of visual space, all arriving at the same neuron in V2.[11] So if you were a neuron getting inputs that signaled the presence of edges at particular angles in bits of the visual world that are next door to each other, what would you like? Exactly: you would like conjoined edges. Which also means each V2 neuron sees a bigger chunk of visual space than any neuron we passed in V1, because it is integrating inputs from V1 neurons that look at different bits of that space.

Seeing a bigger chunk of visual space means that many neurons in V2 know about something V1 neurons are blind to: texture. We can divide the visual world into "things" and "stuff": objects defined by edges, and the stuff that fills in between those edges.[12] The glistening of a field of snow, the leather of a creaky armchair, the cold steel of a knife blade— the crumbliness of a cookie top—we can see the stuff between the edges of the field, of the chair, of the blade, of the cookie. Such textures are patterns of variations in how light is reflected, patches of lighter and darker regions, like the undulating corrugations of the cardboard lid of the cookie box. We already know from our sojourn in V1 that where a lighter patch and a darker patch meet is an edge. A texture is then a particular dense combination of edges of different angles and lengths and thicknesses occurring together. Individual neurons in V1 will see those edges, each neuron sending spikes if the pixel they can see contains their particular preferred angle, length, and thickness of edge. And these V1 neurons will make up the legion of inputs to a single V2 neuron, a V2 neuron whose spikes will thus send a message that a particular dense combination of edges of different angles, lengths, and thicknesses is out there in the world.[13] About texture; about stuff.

Around us are V2 neurons sending spikes to the ragged cardboard of the box lid, the smooth laminate of the desktop, the grainy black plastic of the back of the monitor opposite you. Indeed, these first neurons in V2 encapsulate a common theme of our journey through the visual parts of cortex: that at each stop, neurons combine and transform inputs from the area we just left, creating more complex representations of the world.

This is made abundantly clear when, having repeated our trip through the layers of V2 and back down into the white matter, we shoot into the next main region of the visual cortex, V4 (yes, there is a V3; no, we don't know what it does). The neurons we run into here like clear contrasts between the foreground and the background: they like colors.[14]

We leap a gap onto a neuron crucial to your mission of finding a rapid premeeting pick-me-up. A neuron whose input means it likes the light oaty brown of the cookie surface contrasted against the dull brown of its box. A neuron whose messages will be crucial to identifying that collection of edges, curves, and contrasts as the delicious, delicate morsel you seek.

To understand how this neuron knows about brown all of a sudden we need to back up to the start of the visual system, to the cones in the retina who paused their release of messenger molecules when they absorbed the photons bouncing off the cookie or the box or the desk, and kicked off this whole charade. For as you may know, cones come in three types corresponding to the three wavelengths of light they prefer, which for convenience we'll call "red," "green," and "blue" (more accurately, because "red" just happens to be the name we give that wavelength of light in English, we should call the "red" cone the "long wavelength-tuned cone," but that's as tedious to type as it is to read. So "red" it is. Oh, and that wavelength is not the color red anyway, as you're about to find out).

The three types of cones each kick off their own pathway through the retina and out to the brain via the ganglion cells. With each of the three pathways having all that other retina-dispatched information too: "red"

ON and OFF responses, "red" dim or bright—"red" for all the channels. And neurons in V1 respond to combined versions of these three cone pathways too.[15] So some V1 neurons send spikes to "blue" alone, some to the sum of "red" and "green," some to the difference between "red" and "green," and many others to some mixture of those three things.

But at this point, these are not colors as we know them. They are just responses to the wavelengths of the light reflected from an object, and will be there whether the actual color of the object is white, red, or a shocking pink. Yes, "blue" neurons in V1 will send spikes if Janice's pink-haired troll key ring falls into their bit of visual space, because natural light contains all wavelengths—"red," "green," and "blue"—and so some light of all wavelengths will be reflected into the eye. The key is how much is reflected at each wavelength.[16] The light falling on an object contains a particular mixture of "red," "green," and "blue." Colors are the proportion of that mixture reflected into the eye: if the mixture of light is heavy on the "blue," as it is under the flickering lights of your office, but the reflected light contains proportionally more "red" than in that mixture, then we see red.

(This idea is worth repeating: color is not the wavelength of visible light. Color is the proportion of reflected light at that wavelength compared to the total amount of light at that wavelength. Don't worry, Newton thought it was wavelength too, and he was wrong. As did I, but you don't care what I think.)

It's the neurons in V4 that work all this out. We've arrived as part of a legion of spikes coming from neurons in V1 (and V2) that each respond to one of those cone-pathway combinations. How many spikes they each send signals how much of their cone-pathway combination is present: some will be sending weak signals, a few spikes, because their cone-pathway combination is picking up little reflected light; some will send strong signals, many spikes, because their cone-pathway combination is picking up a lot of reflected light. By receiving two or more of these cone-pathways, the neurons in V4 get to contrast their signals and determine "color": what combinations of wavelengths are strongly reflected and which weakly. The V4 neuron we landed on, the one that likes the oaty-brown of the cookie, must contrast all three pathways: the heady mix of 74 percent reflected "red," 55 percent reflected "green," and 38 percent reflected "blue," that is, approximately, oaty-brown.

We've barely time to register this before we have to grab the next spike out of this V4 neuron, shoot along yet another axon, clones flying off in all directions at each branch, repeat the loop through the layers, and plunge back into the white matter.

To reemerge in the midst of the temporal lobe, among the mass of cortical tissue that cares about shapes.[17] Our spike is part of the legion descending on a pyramidal neuron that will put together the messages we and others are bringing about parts of a long, jagged curve in the space in front of us into the elliptical shape of a cookie top. A truncated ellipse obscured by the straight edge of the tray of the box—another shape put together by nearby neurons who ultimately got their information from bits of the retina just below where we started. Millimeters away, other neurons are putting together their legion of inputs into a rectangular box lid. It's all falling into place: the oaty-brown cookie, the dark brown box, the rosewood desk. You're also just noticing how tediously beige the office colors are.

And just down the temporal lobe from where we've landed are neurons that deal with the shapes humans ultimately care most about: faces. And which you also care about deeply right now, to answer the crucial question: are any faces looking at me while I contemplate this cookie? From V2, via V1, the simple but unique conjoined edges of a nose, a brow, a chin, the line of a mouth, the curve of the cheek bones. Combine those with the colors from V4 of the watery pink lips within graying stubble against pale skin, and you get: a Graham. Facing side-on, across the far side of the office, eyes pointed ceiling-ward, idly contemplating existence.

It would be remiss of me not to point out that our spike's journey is also crucial for a set of shapes that only humans care about: writing. Writing is just edges and lines, corners and angles. The visual system gets very excited by writing. In V4, a bunch of neurons a few millimeters away from where we left were creating their own spikes to signal a set of edges laid out contiguously in a curve that ends roughly in line with its start point. That were responding to the black felt-tip smear of the upside-down "C" of "Cookies" contrasted against the dull brown lid. Near them, neurons loving edges in continuous curves that start and end in the same place: the "o"s. Others salivating over hard straight edges with two short lines kicking out up and down from it at sharp angles—the all-important "k." Reading is primarily an exercise for the visual system, in piecing together a welter of edges and curves and angles into distinct shapes, and those shapes into combinations of shapes—words—and spikes upon spikes while you read this sentence right now.

Highway Do

Back at V1 we missed a turnoff. As the axon of the V1 neuron branched, we followed one cloned spike down into the white matter and off to a neuron in V2 at the start of Highway What. But cloning ourselves along with the spike and taking another branch would have landed us on a V1 neuron at the start of Highway Do.

Unique to Highway Do is the computation of motion. Some simple and complex-like neurons in V1 know about local motion.[18] Just like the rest, they each have their preference for specific angle(s) of edges of a specific fatness in a specific location of the visual world. But they send the most spikes when that lot is combined in an edge that is moving, moving along an axis at 90 degrees to the angle they prefer. So if such a direction-selective V1 neuron likes edges at 45 degrees, then it will send most spikes when that 45-degree edge is coming from top left to bottom right of the tiny pixel of the world that V1 neuron can see. And these direction-selective neurons send their axons onward to start Highway Do.

Flowing out from V1, Highway Do also runs through V2 and V4, through different neurons to the ones we visited in Highway What— different neurons whose roles are, to be frank, not well understood. (We can make some educated guesses: given V2 neurons like conjoined edges, it would be logical if the V2 neurons on Highway Do liked conjoined edges moving in a specific direction.) But the first unique stop on Highway Do is also the one we know most about: V5—or area MT, to its friends.

Area MT neurons put the whole picture together; they respond to global motion, across the whole field of view. Some MT neurons respond to a coherent collection of edges and surfaces moving from left to right. Some to such a collection moving from bottom to top. This sensitivity of an MT neuron to global motion in a particular direction likely comes from their integrating the spikes about local directions pouring into area MT from V1 (and V2) neurons.[19] Imagine we sit on one MT neuron and look at the legion of direction-selective V1 neurons bombarding it with spikes. That legion will contain V1 neurons that like all possible different directions, and each V1 neuron will be looking at a tiny pixel of world. Collectively, the legion of inputs will together cover all directions of possible motion in a big patch of the visual world. So for an MT neuron to like, say, left-to-right global motion, it just has to weight heavily those inputs coming from the many V1 neurons signaling edges approximately moving left to right in their own little patch of space. And voilà: one neuron knows about a coherent collection of edges and surfaces moving in the same direction.

Vitally for you, the neurons in area MT are not sending spikes in response to the coherent collection of edges and angles that make up the cookie. The cookie is not moving. But many other things in the office are. Some neurons in area MT are sending spikes in response to the coherent collection of edges and surfaces you call "Sarah" striding across the office, hair tucked behind her right ear, as she moves across your field of view from left to right. She's close. Too close. You need to know: is she heading toward you, to foil your cookie-snaffling chances, or away from you?

The next stops on Highway Do solve that problem. We tag along with a cloned spike from area MT as it's fired into the variegated pastures of the parietal cortex, to areas labeled VIP and MST. Neurons there integrate the global motion signals along different axes to signal what is happening to the moving object. Is it constantly shifting its axis of motion? Then it's rotating. Is it moving along the top-to-bottom axis but taking up more or less of the visual world? Then it's expanding, or it's contracting. You want a contracting Sarah—let's check out a contracting neuron. Our spike flashes past one and we jump the gap with bated breath, then watch as the multitude of voltage blips accompanying our blip drive a stream of spikes down its axon, meaning: Sarah is moving away from you, heading for the conference room—armed with her coffee-brimmed "It's only feckin Monday again" mug, chimes in Highway What. Your relief is palpable. Cookie, you're nearly mine. Ah, but if only it were so simple.

FORTH, AND BACK

Jumping forward as we've done from area to area, your visual system seems arranged in a neat hierarchy, information from the eye ascending area upon area, each creating an ever-more-complex representation of the visual world. There's some truth to this idea. When researchers train deep neural networks on huge image banks of objects, their simple pretend neurons arranged in rigid layers feeding only forward to the next one, they find their networks recapitulate the rough outline of Highway What's hierarchy.[20] The first layers are like V1, responding to simple edges, with increasingly complex responses across farther layers, reproducing the increasingly complex responses to objects across V1 to V2 to V4.[21] With enough layers, such deep networks can even reproduce the ability of temporal lobe neurons to send spikes about specific shapes, and the better the deep network matches human ability to recognize a pictured object as the correct type—as a car, or a chair, or a table—the better the output layer of the network resembles the activity of temporal lobe neurons.[22] When traveling forward from the eye, thinking of the Highways as an approximate hierarchy is a useful rule of thumb, a schematic guide to making sense of the deep complexity of the visual brain.[23]

There's no doubt though that the visual system is not a strict hierarchy, and no doubt that the two Highways are intertwined. We know they converge in the far distant regions of the cortex from V1, like dorsolateral prefrontal cortex, which we will visit later in chapter 7. There is clearly cross-talk between Highways, allowing spikes from one to influence neurons in the other.[24] And most crucially, there is feedback throughout, from V2 to V1, V4 to V2, everywhere.[25] As we clung to our spike heading forward from the eye through V1 and onward, endless spikes zipped past us the other way, heading back to V2 or back to V1. As though it was necessary for first visual regions to know what farther ones along were already thinking. Which we'll see in chapter 10 might be true.

Even within the hierarchy, digging down reveals nuances. The broad strokes of Highway Do and Highway What trace connections across great swaths of cortex. But at the single neuron level there is likely a strict logic to each neuron's projections, as they target specific types of neurons in specific bits of the hundreds of millions of neurons we call V2 or V4. We've just started uncovering this logic for V1.[26] Using some crazy new tech: RNA barcoding.

Take a unique strand of synthesized RNA. Inject it into a neuron in V1. Then wait. That RNA strand will be transported along that neuron's axon, to everywhere that axon goes. To find out if it goes to area X, cut out that area and sequence the tissue for that unique strand of RNA. If it's there: bingo, your neuron connects to area X. The brilliance of this approach is that you can inject as many unique strands into as many neurons as you have the time and energy to handle, and then sequence for them all in the tissue you cut out. The result is recovery of the precise connections of hundreds of single neurons, a scale once thought unfeasible.

And in the V1 of a mouse, those hundreds of neurons revealed that, indeed, there is a tight logic. Mrsic-Flogel's team (yes, them again) injected barcoded RNA into the neurons of V1 in mice, then sequenced the barcoded RNA in six regions of visual cortex that were potential target areas of V1. Half the V1 neurons targeted two or more of the six areas, but those areas were not random combinations, each neuron did not seemingly pull its combination of target areas out of a hat. Instead, of the sixteen possible combinations of two, three, or four targets, just four combinations dominated. Half the neurons in V1 thus fall into one of four groups according to the combination of neighboring cortical regions where they send their signals. Yet this is but a tantalizing taste of the discoveries to come. We know much about how cortical neurons connect nearby, less about how they connect far—and next to nothing about how they connect to the other side of the brain.

THE OTHER SIDE

When we exited V1 and dived into the white matter, one branch shot off, taking a cloned spike with it, to cross the brain. This branch formed one of the billions of axons of the corpus callosum, the web of wires that link the left and right cortex. If we could clone ourselves, we would have followed that spike too, along that cable, an envoy now, carrying a message to sustain a fragile peace between the left and right brain.

Problem is, I've no idea where we would land. As far as cortical cartographers know, we would almost certainly land in the same region of the cortex on the other side. From V1 on the left to V1 on the right. From left V2 to right V2. But to other regions that deal with vision? Presumably. Anywhere else? We've no idea. For we know little about the journey of spikes between the halves of the cortex. One reason is simply because the technology has not existed to record single spikes from many single neurons in different regions of cortex at the same time. Which includes regions on two sides of the cortex. Another is that dealing with one side of the brain alone is mind-bendingly complex, with an infinite amount to learn, so dealing with two sides is worse.

Technology is catching up. When we zoom out our microscopes and image activity of large regions of both sides of a mouse's cortex at the same time, we see activity is usually synchronized between the same region on the two sides of the brain.[27]

Not just passively synchronized, but seemingly synchronizing each other. In 2016 scientists in the labs of Shaul Druckmann and Karel Svoboda imaged many neurons from the same small region on both sides of the cortex at the same time.[28] When the spikes were switched off in the region on one side, the two sides desynchronized, running free; remarkably, when the spikes were switched back on again, the silenced side immediately caught up with exactly what was going on the other side of the cortex. Spikes crossing between the hemispheres are thus crucial for synchronizing the same region on both sides of the cortex.

This cross-brain traffic of spikes is perhaps more crucial in humans than in any other species, because our brains seem to be the most lateralized of any species; for many of the jobs carried out by our cortex, one side does more of the heavy lifting. The classic example is handedness. If you're right-handed then the bits of hand-controlling cortex on your brain's left side are doing more of the work than those on the right side; if you're left-handed, it's the hand-controlling cortex on your right side that's doing more work.[29] At its most extreme, functions exist in one side of your cortex that are barely detectable in the other. Both the understanding and creation of speech are crucially dependent on regions in your left cortex that do not exist on the right (in almost, but not quite, everyone—strong left-handers tend to have speech regions in the right cortex, or even on both sides[30]). More subtle lateralization of a plethora of functions is evident in functional imaging of human brains. During tasks ranging from calculations to assessing faces, one side of the brain is demanding far more oxygen-rich blood, its neurons yelling "feed me!" the loudest, and thus seems to be doing the heaviest computational lifting during that task.[31]

All this means that you and I (and even Graham) may well depend on our cross-brain spikes to let one side of cortex know the result of computations happening on the other side. Without them, things can get a bit weird.

Quite how weird is brought home by the rare cases of split-brain patients. These fascinating souls had such severe epilepsy that surgeons resorted to cutting the bundle of fibers, the corpus callosum, connecting both sides of the cortex in order to stem the tidal waves of spikes. It worked. But it meant the two sides of the cortex could no longer talk directly.

Working with these patients since the 1970s, Michael Gazzaniga, his colleagues, and others have documented how cutting the communication lines reveals separate functions of the left and right cortex.[32] They do this by simply showing things to the left and right cortex separately: show a picture to the left cortex by putting it in the right visual field; show it to the right cortex by putting it in the left visual field. And from the differences in response between the left and right cortex, we can infer what spikes traveling between them must be coordinating.

One striking difference is in problem solving. In one study, split-brain patients were asked to predict which of two light bulbs would flash on next by pressing a button corresponding to that bulb. One bulb flashed 80 percent of the time, the other 20 percent. People normally match these probabilities, so after a long run of bulb flashes, they will tend to press one button 80 percent of the time and the other 20 percent. Showing this problem to the left cortex of split-brain patients elicited exactly this matching. But showing it only to the right cortex produced maxing: the patient consistently pressed the button corresponding to the most likely bulb to flash (which, incidentally, is the better solution: always pressing the most likely button gets you a guaranteed 80 percent correct; matching probabilities will almost certainly do worse). The two halves of cortex inferred different solutions to the same information. Which means spikes between the left and right cortex are needed to inform each other about their own solutions—and use one.

As you might suspect from the extreme lateralization of language, split-brain patients show us how cross-cortex spikes coordinate vision and words. Shown a fork so that only the right side of the brain can see it, split-brain patients cannot name it. Yet when the fork is passed to their left hand, also controlled by the right brain, the patients can use it just fine. Their right cortex knows what it is but can't access the words, because words are only accessible in the left cortex.

Perhaps the most vital job of the cross-cortex spikes is for the right cortex to tell the left cortex what the hell just happened. Because of the extreme lateralization of language on the left cortex, actions taken by the right cortex in split-brain patients are not accessible to the left cortex speech centers. But the left cortex interprets the action anyway—wrongly. Like in Gazzaniga's favorite story of the claw and the snow.

A patient was shown a chicken claw to the left cortex and a snow scene to the right cortex, and each side could also see its own set of four objects on the table. The patient was asked to point with each hand to the most relevant object for the picture. The left hand pointed to a shovel, fittingly for the right cortex's view of the snow; the right hand pointed to a chicken, a sensible match for the left cortex's view of a chicken's claw. But, of course, only the left cortex has access to language. So when the patient was asked why those two objects had been chosen, the patient responded, "The chicken claw goes with the chicken . . ."

(left cortex language centers seemingly knowing about both the picture shown to its visual parts and the object pointed to by its motor parts); the patient went on ". . . and you need a shovel to clean out the chicken shed." Left cortex seemingly caught in the act of making stuff up, as it had no idea why right cortex picked a shovel—the language centers in left cortex knowing nothing of the snow scene.

So more than just matching words to objects, how we interpret the world vitally depends on our cross-cortex spikes. For the right side of your cortex would like to know what the left side is saying, both to itself and to the right side of your body. And the left side of the cortex is equally keen to hear back from the right side. It is those spikes flowing across the corpus callosum that seem crucial to making two halves of cortex into your one interpretable experience of your own body.

Time to move on. In much less than a second, indeed in just a few hundred milliseconds, we've tracked a succession of cloned spikes across the visual regions of the cortex, across and between both sides of the brain. We've spun through connections within layers to next door neurons, up and down the layers within a region of cortex, plunging into the white matter, only to be fired up again into a new region, a new local circuit to traverse. Those few hundred milliseconds were enough to transform spikes from the retina about elementary pixels into spikes about fully fledged representations of a cookie, a box, a desk, the people around you, and where they are. Time to decide what to do about that cookie. And here we make our first bad decision.

CHAPTER 5

Failure

WHAT IT MEANS TO FAIL

So many clones to choose from. Pick one. Flying out from the visual regions into the central regions of the cortex, our journey comes to a screeching, sudden halt. Our spike reached a synapse on the axon, and nothing happened. No bags of chemicals were dumped out. Nothing tickled the neuron on the other side. The spike's message is lost forever to the other neuron. It failed.

We stare at evolution in slight disbelief. It took an extraordinary effort to make that spike, a legion of fellow spikes arriving together, at the right place and time on the tree of one neuron, and a glut of energy to drive the openings and closings that create the all-or-nothing pulse of electricity. Yet for all that, it failed. Information has been irretrievably lost.[1] What kind of inept cowboy threw this bag of cells together and called it a "brain"?

Spike failure is a bug, a flaw, a potentially inevitable consequence of running up against the limits of making things work in biology. On such a microscopic scale, where a neuron's body is tenfold smaller than the width of a human hair, noise comes from everywhere, from tiny changes in temperature to tiny movements of the brain, all too small to be noticed by something as big as a fly, or a mouse, or a human. So when the spike arrived, the chain of events that leads to the dumping of the bags of chemicals was disrupted by noise. And sometimes the bags of chemicals have just run out, so when our spike arrived there was just nothing left to give.

But rates of failure differ wildly between different parts of the brain, and even between different types of neuron in the same part of the brain.[2] The spikes from some neurons fail at an alarming rate: at excitatory synapses in the hippocampus, about 70 percent of spikes never make anything happen; at the worst offenders, this figure is 95 percent.[3] Ninety-five percent. Only 5 percent of all spikes arriving at those synapses create a voltage blip in the neuron on the other side.

Yet other synapses in the brain have a failure rate of zero. Every spike arriving creates a response on the other side.[4] Even weirder, different synapses between the same pair of neurons can have dramatically different rates of failure.[5] If failure was a bug, then it makes no sense that it can differ so radically. Then perhaps failure is a feature, not a bug.

Indeed, this small but persistent annoyance is actually a powerful computational tool.

WHY FAIL—TO COMMUNICATE BETTER

It may not surprise you to learn that theorists love spike failure too. It's another one of those weird paradoxes. The brain uses spikes to send information between neurons, yet by having synaptic failure the brain deliberately prevents itself from doing so. Why would the brain prevent itself from sending information? Many ideas have sprung forth from the minds of theorists as they rub their hands together with glee.

One simple idea is that spike failure is necessary to control how strongly neurons connect to one another. You may recall that the legion of inputs to a neuron have different strengths, some are weak and some are strong. Having run into a failed synapse, we can now see that the thing I called strength is made up of two parts. The first is the size of the voltage blip; the second is the reliability of the synapse.[6] An input at one synapse may be able to make a large blip, but if it does so for only 10 percent of the arriving spikes, then that input weakly influences the target neuron. Theorists have long pointed out that this gives two options for changing the strength of a connection from one neuron to another: we can either change the size of the response or change its reliability.[7]

Changing the size of the response is tough; it means increasing the number of packets of molecules at the synapse, increasing the number of receptors on the other side, or both. Changing the reliability offers a different option; it would just require making the release side more or less sensitive to each spike. Indeed, in experiments on connections between neurons of the hippocampus, Charles Stevens and Yanyan Wang showed that repeatedly stimulating a pair of connected neurons can turn their unreliable connection into a reliable one.[8] And as you can't make a reliable synapse any more reliable, most synapses have to start off unreliable to allow headroom for change. Which all means something remarkable for learning. As many forms of learning are thought to depend on changes to the strengths of connections between neurons, this theory implies the brain is deliberately noisy in order to allow learning. Learning becomes the routing of reliable spikes, the transmission of reliable information between neurons.

William Levy and Robert Baxter took on the idea that unreliable synapses literally lose information, to ask: what if this loss is deliberate?[9] They propose that unreliable synapses are actually a very clever solution to the fundamental problem of how to transmit as much information as possible while keeping energy costs as low as possible. Sending each input into a neuron takes energy. Lots of energy. Sucking back up the released molecules for reuse, packaging them back into bags, and of course making the voltage blip itself, with the opening and closing of holes in the neuron's skin. In fact, synapses and their machinery take up about 56 percent of all the energy used by neurons in the brain.[10] And as the brain uses 20 percent of all your energy, so reducing those expensive synapses will give you more energy for the essentials of life.

Levy and Baxter pointed out that the axon of a pyramidal neuron has a limited capacity to transmit information, which is set by how many spikes it sends on average per second. For a neuron to be energy efficient it needs at least this much information in its input, so that the output is not wasting energy sending nothing. But we already know that a pyramidal neuron gets about 7,500 inputs. If even just 10 percent of these inputs were transmitting every spike across their synapse, the total information input to a pyramidal neuron would vastly exceed the capacity of its axon to output information. The neuron would be overwhelmed with more information than it could use, so most of those spikes coming across synapses would be unused information, wasting energy.

In Levy and Baxter's "matching" theory, synaptic failure comes to the rescue. By preventing spikes from sending information to the target neuron, failure brings down the information rate of the neuron's input. With a high enough failure rate, it can bring down the input's information rate to match the capacity of the neuron's output axon. And in doing so, it would perfectly balance the amount of energy used, to use all of the axon's capacity to send information, but no more. Beautifully, it turns out that the failure rate in this theory should be around 75 percent: exactly what we see in the cortex.

Harris, Jolivet, and Attwell pointed out that the same basic idea—conserve energy—gives us another reason for synaptic failure.[11] When one neuron connects to another it often makes multiple contacts—multiple synapses clustered together on a bit of dendrite. If all those contacts relayed every spike faithfully, then the target neuron would be seeing the same spike multiple times. The information coming from the sending neuron would be duplicated, making some of those spikes redundant. And redundancy wastes energy: sending the same spike four times to the same neuron costs four times the energy for no gain in information.[12]

Synaptic failure comes to the rescue here too. If those multiple contacts were each unreliable, then the brain could ensure that most of those contacts failed to transmit the same spike. Indeed, the failure rate could be set so that at most one of those contacts transmitted the spike (on average), eliminating the redundant information. In which case, the brain would again be maximizing the amount of information transmitted for the least energy.

This "redundancy" theory makes a strange prediction: the more contacts a neuron makes on a target neuron, the more unreliable those synapses should be. That prediction is true. Studying pairs of neurons from the hippocampus, Tiago Branco and colleagues at University College London showed precisely that the more synapses one neuron makes on another, the higher the failure rate of those synapses.[13]

Both the "matching" and "redundancy" theories also predict that synaptic failure must be under the control of the target neuron. Both theories propose that synaptic failure exists to throttle information coming in to a neuron in order to increase the energy efficiency of synapses. But only the target neuron knows which direction to change the synaptic failure: it knows the capacity of its axon, and it knows how much redundancy exists on its own tree. Branco and colleagues showed the failure rate was indeed under the control of the target neuron. When they stimulated a local bit of the dendritic tree, as though that tree was getting more inputs, the failure rate of the synapses on that bit increased. Synaptic failure could thus be a neuron's greatest life hack, an all-purpose tool for optimizing the efficiency of energy and communication.

WHY FAIL-TO COMPUTE MORE

Synaptic failure could do more than control how well neurons communicate. It could also create new ways for them to compute. For failure allows neurons to do cool things with spikes arriving close together in time at the same synapse. And that's because failure can change the apparent strength of synapses within a few milliseconds without altering anything about the synapse at all (figure 5.1).[14]

Imagine two spikes arriving in short order at a synapse. If that synapse has a low failure rate for the first spike, it is less likely to transmit the second. Why? Because the first spike used up lots of the bags of molecules that are ready to be dumped on other side; so with a reliable first spike, a second spike in rapid succession may deplete the bags too much. The synapses need time to recover. Indeed, enough spikes in a row can completely empty the bags, needing a long time—about 10 seconds—to recharge.[15] So the brain uses failure to throttle the rate of release to sequential spikes.[16] Which means the strength of the synapse gets progressively weaker to later spikes in a sequence. The synapse has what we call short-term depression.

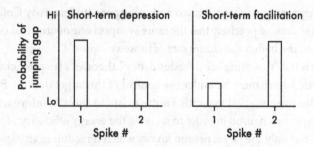

FIGURE 5.1. Short-term depression and short-term facilitation. We imagine two spikes arriving at a single gap in quick succession; we plot the probability of each spike causing a release of molecules across the gap. The probability decreases if the synapse is depressing (*left*) and increases if the synapse is facilitating (*right*).

Conversely, when a synapse has a high failure rate for the first spike, it is more likely to transmit the second. Why? Because the first spike primed the synapse to release the bags of molecules if another spike arrives quickly thereafter (where quickly is a hundred milliseconds or less). Which means the strength of the synapse gets progressively stronger to later spikes in a sequence. The synapse has short-term facilitation.

To us, this is great news. Our spike failed, but if we linger here a few moments more, another spike clone will arrive and immediately trigger the dumping and diffusion of molecules, and the evocation of the voltage blip, for us to follow. As we wait to continue our journey, I'll reel off our theories for how these short-term changes in strength create new forms of computation.

For one, equipping a synapse with short-term changes means the receiving neuron can compute things based on sequences of spikes. Here's an example from Wolfgang Maass and Tony Zador.[17] Consider a neuron with the following problem: if two spikes arrive at one of its synapses within a few milliseconds of each other, then that neuron needs to know; anything farther apart it can ignore. For example, a neuron in a small, furry, scared rodent, in whose world two rapid hoots in succession indicate an imminent owl. Two rapid hoots creates two rapid spikes, and the alarm needs sounding. Something hoot-like but farther apart also creates two spikes, but is not an owl; it's a false alarm, so needs ignoring.

Now have those two spikes arrive at a synapse with short-term facilitation. If two spikes arrive farther apart than X milliseconds, then both the first and second spikes fail, and the neuron knows nothing about them. False alarm, no reaction. But if the two spikes arrive within X milliseconds, then the first fails but the second is transmitted. Equipped with short-term facilitation, the receiving neuron knows that two spikes occurred within X milliseconds, even though the first spike never arrived. Awogah awogah, hurry scurry our furry friend. In effect, synaptic failure allows the brain to compute with ghosts, spikes that were never there.

Better still, short-term depression is a filter. A filter that allows neurons to ignore oscillations in their inputs. There are many states of the brain in which some neurons oscillate their outputs—their rate of spikes goes up and down regularly over time. Deep sleep, for example, has many neurons in the cortex jump together between a burst of silence and a burst of spikes every few seconds. Any such rhythm means there are periods when spikes occur close together. But if those spikes arrive at a synapse equipped with short-term depression, then the rhythm will be broken. For the synapse will likely transmit the first spike, but less likely to respond to a second, and even less to a third, and so on. Even better, just having a high rate of synaptic failure will act as a filter, for it will randomly lose spikes all through their ebb and flow of the incoming oscillation. To the receiving neuron, the rhythm is lost, filtered out.[18]

Our most effective treatment for Parkinson's disease might depend on just this filtering effect of synaptic failure. Deep in a brain with Parkinson's disease lie groups of neurons that persistently oscillate when they should not. These neurons in turn entrain their target neurons to the same rhythm. And this entraining seems to disrupt movement signals in both the cortex and brain stem, thus creating the characteristic difficulties with moving seen in Parkinson's.

Remember deep brain stimulation? This delivers a constant, clockwork stream of electrical pulses directly into the subthalamic nucleus, a core group of those persistently oscillating neurons. This clocklike rhythm gives back to Parkinson's patients control of their own movements. But how is a mystery.[19] On the face of it, deep brain stimulation would seem to overwrite one form of unnatural neural activity, the persistent oscillations, with another, the clockwork tick of its electrical pulses.

A compelling theory is that deep brain stimulation's therapeutic success can be explained by synaptic failure. This constant stream of pulses indeed dramatically increases the number of spikes being sent down the axons of those core neurons targeted by the stimulation. But this dramatic increase in spikes along the axon in turn increases the rate of synaptic failure when those spikes arrive. And what does lots of synaptic failure do? Filters out oscillations! Thus, the theory goes, the receiving neurons do not see the persistent oscillations anymore and can respond freely, normally, restoring the control of movement.[20]

This is one of those fantastic coming-togethers of science, a case study in why it takes all kinds of knowledge and all types of research to make progress. The experimental facts that synaptic failure exists and how it works were established in slices of rat brain, for the pursuit of pure science, to simply understand better how the brain works. Theorists fascinated by this weirdness wanted to understand why synaptic failure exists and how the brain can still work in the face of failure, and they came up with the above theories for how synaptic failure creates new ways of communicating and computing with spikes—including filters. Yet other experimentalists established that a core group of neurons have persistent oscillations during Parkinson's disease; yet others, what effect deep brain stimulation has on its target neurons. Finally, a group of theorists and experimentalists take all those pieces and complete the puzzle, to show that all this work, these decades of basic and clinical science combined, fit together into a deep explanation for how deep brain stimulation works. In science, the answer is never where you're looking for it.

WHY FAIL—TO DECIDE
WHEN TO BE AFRAID

More than just theories, we've even caught the new computations enabled by synaptic failure in action. Startling work from Tiago Branco's lab at University College London has shown us that evolution has co-opted failure to define the threshold for being afraid enough to flee.[21]

How do we know when to flee? Remember last Thursday. Your report was late. Through the crowd of bodies milling about the office you glimpsed the boss's glass door opening, then a flash of puce that could be the boss's Thursday tie, a snatch of voices saying the boss's name, a definite glimpse of absurdly expensive Italian leather striding toward your desk. Time to go, exit stage left. Each new thing you notice added more evidence of an imminent threat. And when enough evidence was gathered, when the cumulative creeping sensation got too much, your brain decided the threat was real—and escape was the best course of action.

Somewhere in your brain adds up evidence of a threat, and somewhere then sets a threshold for when enough evidence means "Flee!" To find out where, we can't sit people in a room and menace them with disgruntled bosses, lions, or clowns. And even if we were allowed to, we can't stick electrodes inside their brains to record their neurons while being clown-menaced. So Branco and team turned to that workhorse of neuroscience, the mouse.

Place a mouse in a box, with a handy little dark shelter at one end to feel safe in. Let it roam free, exploring, scouting its new home. Now make a shadow fall over it, a shadow that rapidly grows in size. A rapidly growing shadow that looks uncannily like a diving bird. Result: mouse charges back to the shelter and cowers deep in the darkness, little heart thudding.

The clever bit is that the darkness of the shadow is the evidence of an imminent threat: the darker the shadow, the greater the perceived threat. If the shadow is very dark, the mouse runs as soon as it starts to get bigger. If the shadow is very light, it can take four or five repeats of the swooping shadow for the mouse to finally decide to run for shelter.[22]

With this swooping shadow controlling both evidence and escape, we can then look in the brain and ask: what adds up the evidence of a swooping shadow, and what drives "leg it" when that evidence hits a threshold? Branco and team already had clues about where to start; for we already know that it should somehow involve the superior colliculus.

The superior colliculus sits atop the brain stem and is the privileged recipient of information directly from the retina—it gets to know what's going on in the world long before you do. (Before you ask: yes, there is also an "inferior" colliculus. And as the superior colliculus does seeing, and the inferior does hearing, that gives you some idea of the pecking order in neuroscience.) And the output of the superior colliculus definitely controls movement. So if we're looking for a bit of brain that can quickly add up information from the eyes and then make you move, the superior colliculus is the superior candidate.

So Branco and team used the armory of modern neuroscience to find out if this was more than mere guesswork.

They recorded the activity of the output neurons of the superior colliculus. The neurons dramatically increased their activity right from the start of the swooping shadow. The activity increased far more when escape happened than when it didn't. And the stronger the activity, the faster the mouse started to escape. As though the activity in superior colliculus was adding up evidence of a threat.

They switched off those output neurons. Now the mouse did not react to the swooping shadow, just carried on exploring. As though someone had completely removed its threat detector.

They turned on these output neurons of the superior colliculus when there was no swooping shadow. This was the crucial test, to see if the brain could be fooled into thinking there was an imminent threat. And it could: turning on these neurons made the mouse run for shelter. Even better, the more the neurons were activated, the more likely the mouse ran to the shelter, exactly as though the neurons were signaling the imminence of danger.

All in all, pretty damn convincing that the superior colliculus is the threat detector. The obvious next question was: what is the superior colliculus talking to, saying "it's time to run"?

Enter the periaqueductal gray. Good grief, neuroscientists are so bad at names: it's a bit of gray stuff around ("peri") an aqueduct in the brain. Let's call it PeGy, shall we? PeGy does a lot of things. One bit of it controls urination, for example. But it gets a massive input from the superior colliculus, and it controls a lot of rapid reactions.

So Branco and team dusted off their neuroscience magic tricks again, popped some new mice in the swooping-shadow box, and set to work.

They turned off PeGy's neurons. Now the mice did see the swooping shadow, did react to the threat, but they didn't run away. At all. They just froze in place (the other option: stop moving, as predator brains track movement). PeGy, it seems, controls running away.

They recorded PeGy's neurons. Their activity increased right as the escape started, and not a moment before. So PeGy's neurons don't add up evidence, but sure seem to mean "run."

And the clincher: they turned on PeGy's neurons when no shadow was swooping, again pretending there was a threat. Beautifully, as they activated more and more neurons, they found an all-or-nothing response. If too few neurons were activated, the mouse never ran for shelter. But if just enough neurons were activated, the mouse always ran for shelter. Nothing in between—no ifs, buts, or maybes. When PeGy says go, you go.

That all-or-nothing response raises the crucial question: what sets that threshold between escaping or not? It's definitely something between the colliculus and PeGy, for when Branco and team turned off just that connection it turned off escaping completely. It seems the superior colliculus activity goes straight into PeGy's neurons to turn evidence of a threat into running away. So why doesn't every increase in superior colliculus activity trigger PeGy's command to run away?

Because it turns out the connections from the superior colliculus to PeGy are weak, and rubbish. For just like in the cortex, each spike barreling down from one colliculus neuron into PeGy creates just a tiny blip of voltage in the PeGy neuron. And so it would need tens or hundreds of these blips to make a PeGy neuron send just one spike. Each connection is weak. And each connection fails. On average, 20 percent of all spikes from the colliculus do nothing at all in PeGy.

A weak, rubbish connection means that the colliculus needs to send a lot of spikes, all at the same time, to get a response out of PeGy. The weak, rubbishness is the threshold. Those light swooping shadows, each eliciting just a bit of activity from the colliculus, did not create enough activity to overcome the weak rubbishness of the connection to PeGy. But the dark, intense shadows drove a wave of intense activity out of colliculus into PeGy, overwhelmed the weak rubbishness, and escape was the only option.

This work has two particularly beautiful lessons for those who study the brain. The first is that we have a complex behavior in a mammal—escaping by running to a shelter—now drilled down to a single connection in the brain, and the properties of that specific connection. For that connection is the threshold; its weakness and unreliability set the threshold—they are the limits to overcome.

The second is that evolution has co-opted the failure of spikes to create a threshold between escaping and not escaping. Has used the failure of spikes as a way to filter out things that are not threatening, to make sure you don't run away at every sudden noise, or from every looming shadow. And you know what? If you do run away at every sudden noise, or from every looming shadow, this might just mean your connections between the colliculus and PeGy don't fail enough—your brain is just too perfect.

WHY FAIL-TO SOLVE PROBLEMS

I think evolution has co-opted spike failure for far more than just setting a threshold between staying and fleeing. I think synaptic failure is deliberate noise in the brain. Noise on purpose. Evolved noise. And I'm going to propose to you that this noise is crucial to the brain's algorithms for learning and searching.

We know at least two good reasons why noise is good for the artifice brains of artificial intelligence. The first reason is to generalize what you've learned. The second is to search for the best solution to a problem.

Your brain is adept at taking what it's learned and creating general principles. Like your concepts of cars, buses, and gorillas. Having seen some examples of gorillas, your brain can now recognize a gorilla from all sorts of angles you've never seen a gorilla from before, and would never want to if at all possible. Adeptness at generalizing can save you from an ignominious end as a gorilla's cushion.

Artificial neural networks are not adept at generalizing. An AI researcher may train one of their neural networks on tens of thousands of images so that it learns to classify them: "cars," "gorillas," "ice cream vans on fire (irony)." But deep neural networks have many layers of thousands of simple neuron-like units. So they have millions, tens of millions, of connections between those units, and the strength of every connection can be adjusted. Having far more connections to adjust than images to learn means artificial networks are prone to horrible overfitting;[23] they learn the fine detail of each image, become fine-tuned to the nuances. Which means they have not learned the common principles of cars or gorillas or flaming ice cream vans. They can struggle to generalize. Test the trained network on new images of already learned categories—a gorilla, but from the top; an ice cream van gently smoldering—and it fails to put them in the correct categories. Even changing a few pixels of an already learned image can make the network fail.

A widely used solution is DropConnect.[24] Which does exactly what it says on the box: for every new image presented during training, a bunch of the connections in the network are dropped at random. And only the retained connections are updated by the success or failure in categorizing that image. Repeated for each image, this essentially means that every image is presented to a unique version of the network, stopping the whole network being fine-tuned to the details of each image. And when this network is then tested on unseen images, it does a better job of categorizing them correctly. Dropping connections at random adds noise to the network, noise that lets the network generalize.

Your brain faces the same challenges—worse even. Your cortex has billions of connections it could adjust every time you learn something. So how does it not overfit? I suggest synaptic failure. Synaptic failure is exactly the same mechanism as DropConnect: it drops connections between neurons, at random, and temporarily. It adds deliberate noise to the brain exactly where you'd want it to prevent your brain from overfitting. To let it generalize.

The second good reason to add deliberate noise to your brain is to help it search better. Many challenges in machine learning are about finding the optimal solution to a problem given some constraints. Like finding the fastest route between two locations—where fastest often does not mean shortest, and the solutions are constrained by speed limits, traffic, your mode of transport, time of day, likelihood of rampaging sheep escaping a field, and innumerable other factors.

A machine solving these problems will explore the space of possible solutions. It will propose a solution, evaluate how good it is, and work out how to adjust that solution to find a better one.[25] Told you want to go from London to Paris, it will propose one possible route, work out how long it will take, then look at how it can be adjusted—taking a left turn here to get on to a longer road, but one that has a higher speed limit. And repeat this propose-and-adjust cycle until its proposed solution cannot be bettered.

The fundamental challenge for searching is that such problems will have many solutions that are adequate, passable, OK, but which are all surrounded by worse solutions, so that any small adjustments from the adequate solution gets worse results. These are little traps in the space of possible solutions. They make these solutions look like they cannot be bettered—but they can, and there may be vastly better solutions out there, somewhere.

The many ways of escaping these traps boil down to the same thing: add noise. Deliberately making big, random adjustments to a solution lets the machine jump out of traps, to find a better solution, to carry on looking for the best solution. Instead of fiddling with the choice of back roads in Kent and finding no better solution for driving from London to Paris, a big injection of noise could jump the machine's search into the motorway straight from London to the UK's south coast ferry port. (In an ideal world, an even bigger injection of noise would jump the machine's solution to "take the direct London-to-Paris train, you idiot.") Finding solutions by searching needs noise. Even better, it needs noise that can be tuned—made large for big jumps and small to hone in on solutions.

To me, this suggests the tantalizing idea that synaptic failure is at the heart of the brain's search algorithm.[26] Brains need to seek solutions to many constrained problems. Like routes to sources of food that are the best combination of quick, easy, and safe. Just like machines, brains need to traverse the space of possible solutions to find the best (well, more accurately, the least worst). And if this book has made any impression on your brain at all, you know that traversing is done by the sending of spikes between neurons. So to jump out of the traps of adequate solutions the brain needs to add noise to the sending of spikes. Synaptic failure is precisely this noise: random, always there, and tunable—it can be made small or large.

The fact that learning and searching are vastly improved by noise gives us reasons why the brain should be deliberately imperfect, should be full of noise. And, to me, synaptic failure is precisely the form of noise you would want.

Another spike has arrived at the synapse where we've been stranded, a mere few tens of milliseconds after our initial failure. With the synapse primed by that first spike, this second effortlessly triggers the molecular release. We float happily to the other side and follow the voltage blip down the tree. We've arrived at the first neuron in the prefrontal cortex, the great swath of neurons covering the front half of your brain. Dark here, isn't it?

CHAPTER 6

The Dark Neuron Problem

THE CLASSIC VIEW

Every time we've exited the white matter, the view has been awe-inspiring. Like gazing at the night sky, all around us a panorama of scintillating neurons, points of light sparking with spikes. But just like the night sky, there is more darkness than light. Those bright points in the firmament of cortex are but pinpricks on a massive, engulfing darkness. A darkness of neurons that are not firing.

Wherever we follow our spike, there is darkness. Now closing in on the final regions of the cortex that deal with vision, where color is merged with shape ("a light brown, chocolate chip cookie!"), where curves are formed into faces ("Angela and Ishmael are at the office door, not looking this way"), our spike has zipped past billions of neurons in your cortex. The overwhelming majority of them have not fired a spike in the second we have traversed your brain so far. The overwhelming majority do not send anything. Even neuroscientists can find this hard to grasp, and little wonder, as our data have shown the opposite.

Neuroimaging—functional MRI—shows us Technicolor images of the cortex, its regions lit up in a swirling riot of poorly chosen colors that make the Pantone people cry into their tasteful coffee mugs. The swirling colors seem to show us that the cortex is a swarm of activity. That when we see a face the visual areas of our cortex bloom with barrages of neural firing, from V1, V4, and down to the face areas of the temporal lobe. That when we hear a swell of strings, the auditory areas of our cortex bloom with barrages of spikes.

Classic studies of single neurons seem to show us that each and every neuron has a role. That each responds to something—a line, a corner, a movement, a color. For when experimenters lower the thin sliver of an electrode into the cortex, they can readily record the spikes as they are dispatched from a neuron's body. And can relate those spikes to something happening in the world. Many have related those spikes to features of the visual world, like the neurons we have already met, the simple and complex, the edges and lines and angles, the contrasts, shapes, objects, faces. Others we could have met had we taken a different starting point: neurons in auditory cortex that respond to sounds of a particular frequency; neurons in somatosensory cortex that respond to touch on a finger, a toe, an arm.

We've decades of work, tens of thousands of studies, showing us that, when we lower an electrode into a bit of cortex there they are: a swath of neurons responding to the things they like. So surely all neurons send spikes.

Some simple arithmetic will show why this makes no sense. Back when we were basking in the simple cell in V1, if we'd hung out a little longer, we would have seen it fires perhaps 5 spikes every second. And we know it needs about 100 spikes arriving at its excitatory synapses to make one new spike. So 5 spikes in a second means it needs a total of at least 500 excitatory spikes arriving in that second. But we also know that a V1 neuron has about 7,500 excitatory inputs. If each of those inputs was sending at 5 spikes per second, there would be 50,000 input spikes in total every single second. That's too many by a factor of 100.[1] The simple cell in V1 should be sending 500 spikes every second.

But it isn't, and neither are its inputs. And they can't—sending 500 spikes every second is a neuron screaming at the top of its lungs. It's about the theoretical maximum rate a cortical neuron can produce spikes, if forced to do so by an experimenter. Not least because after each spike there is a few milliseconds in which a neuron can't make a new spike at all. No, even the most active neuron in the cortex can only sustain a continuous output of about 30 spikes per second. A paradox: the neurons in cortex are sending spikes at least an order of magnitude lower than they should be if all their inputs were similarly active.

The only way out of this paradox seems to be that most inputs to a neuron in the cortex are not sending spikes. Which implies that most neurons in the cortex are not sending spikes. Is this true?

HOW TO FIND A DARK NEURON

Up until the 1990s, neuroscientists could only record individual neurons in an animal blindly. They inserted their tiny, sharp electrode into some bit of cortex, and only knew they had found a neuron by the blip it made on their oscilloscope, or the noise it made on the speakers in their lab: tick tick tick tick. . . . Which meant they could only find active neurons, because their only way of finding neurons was by their activity.

This created a fearful bias in our understanding of spikes and which neurons sent them. If every neuron you record sends spikes, then as you are sampling at random, blindly, this implies every neuron is sending spikes. But if your only way of finding neurons is by finding spikes, then by definition you could not find neurons that did not send spikes. Indeed they would be dark matter: contributing mass to the brain, but invisible to all your instruments of measurement.

But then neuron imaging came along. We point a digital video camera at a bit of brain, and in that bit of brain each neuron contains a chemical we've injected that lights up when the neuron is active. Most often, that fluorescent chemical is responding to the amount of calcium in the neuron's body, lighting up to an influx of calcium with every spike.[2] By filming the bit of brain in sharp focus, we can see all the neurons with our eyes, see their outlines. And we can see which ones light up. It turns out for decades we've only been recording the tip of the iceberg. Most neurons we can see in these videos are not active.

Our first clue to the scale of silence came from imaging the cortex of anesthetized rats. Under many anesthetics the cortex operates just like it does in deep sleep, jumping between active and quiet periods about once a second. Imaging the first auditory and somatosensory (touch) bits of cortex, Jason Kerr and colleagues reported that the "active" periods actually contained only detectable spikes from just 10 percent of all neurons.[3] Ninety percent were silent in each second-or-longer "active" period, and almost all totally silent in the "quiet period." And this scale of silence wasn't because of the anesthesia. It is found in the cortex of behaving animals too.

Christopher Harvey and colleagues in David Tank's lab at Princeton imaged a section of parietal cortex, which lies at the end of Highway Do, in mice running inside a T-shaped maze.[4] (A T-shaped maze in virtual reality: the mouse ran on a ball, while the virtual world moved around it.) They reported that just 47 percent of neurons were "active" while the mice ran the maze. Even that low number required stretching the definition of active to its breaking point: a neuron was deemed "active" if it had more than two spike-like events in a whole minute, vastly longer, ten times longer, than it took the mouse to run the length of the maze.

And just in case you're thinking that putting mice in a virtual world may muck up their neural activity, rest assured. Researchers in Karel Svoboda's lab have made an industry of imaging neurons in mice doing stuff in the real world. In a study led by Simon Peron, they imaged the specialized bit of the cortex that gets input from a whisker, while that whisker was being used to find a pole (and which the mouse wanted to find, as it was thirsty and the location of the pole told it which spout would have water in it).[5] Even in this special bit of whisker cortex that only cares about that one specific whisker, they found only 67 percent of neurons were active. And again this stretched the definition of "active" to its breaking point, needing one spike event every 100 seconds, ten times longer than the entire sequence of the task. Everywhere we've imaged, most neurons send no spikes in over a whole minute.

Such imaging studies have repeatedly shown silent neurons in the cortex, but left open many problems: is there some technical problem with the fluorescent chemicals we used? Perhaps they did not respond to isolated spikes and so made neurons look more silent then they were. Or did the chemicals not get taken up by all neurons? In which case the "silent" neurons could just be those that had no chemical inside them. Or did they get damaged by the chemicals? In which case, the mere presence of the chemical alone caused neurons to stop spiking. And most imaging studies, including all the above, look at neurons in layers two and three, just at the top of cortex, because it's harder for light to penetrate deeper and thus for us to video deeper. Which leaves open the possibility that there's something special about these layers of cortex, and deeper neurons are all merrily spiking away. As in any area of science, each technology brings great insights but just as many new potential drawbacks. But other teams then proved silent neurons existed using the phenomenally fiddly technique of patch clamping.

Traditionally, neuroscientists just lower a sliver of metal or glass into the brain and pick up spikes when that sliver happens to be near a neuron's body. By contrast, patch clamping finds a neuron within an animal's brain by attempting to physically attach to it—"patching" to it. Because they are finding neurons by physical contact alone, the experimenters are not relying on activity. Patch clamping has its own biases— it is easier to patch a bigger neuron than a small one, and in a live animal you still cannot see what you're doing—but, crucially for us, activity is not one of them. Once attached, they can play the animal some sounds, or get it to touch something, and see if their attached neuron becomes active.

Largely, it does not. Tomáš Hromádka in Tony Zador's lab at Cold Spring Harbor patched a collection of neurons in the first bit of auditory cortex (A1) in awake rats and found most of them were silent most of the time.[6] And silent regardless of whether the animal was sitting quietly or listening to an extremely dull collection of pure tones. Playing sounds to the bit of the cortex that cares most about sounds evoked very little response. Dan O'Connor, then in Karel Svoboda's lab at Janelia Farm, patched a collection of neurons in that specialized whisker bit of cortex in mice, mice that were again using a whisker to find a vertical pole.[7] Guess what? Most of those neurons were silent most of the time. Even when the whisker was waving back and forth, hitting the pole. And both these and other studies have consistently found silent neurons in all layers of the cortex.[8]

Looking back, this epidemic of silence was there in plain sight. Theorists long ago worked out how many neurons should be within recording range of an electrode lowered into the cortex of a rodent. Simple physics says that the farther the distance between the electrode and a neuron, the weaker the spike signal will be. The strength of the signal should drop roughly exponentially—rapidly at first, then slowly—with increasing distance. And there will be some distance beyond which the signal will be too weak to detect with your equipment, because it will be indistinguishable from noise. So theorists imagined sticking an electrode into a collection of neurons packed as tightly as they are in cortex, worked out at what distance the spikes cannot be detected, and counted how many neurons sit within that distance. The answer was: at least one hundred.[9]

But when neuroscientists lower a single electrode into position, they see at most a handful of spikes from different neurons (we can work out they are from different neurons if the spikes are consistently different heights). They see nothing like one hundred neurons. Not even close. So all along that meant: most neurons are silent.[10]

The irony is that dark neurons are essential to being able to record neurons in the cortex at all. If many of those one hundred neurons were spiking, then tens of thousands of experiments would have failed. For the electrode would simply be awash with spikes, a ceaselessly fluctuating voltage, from which individual spikes from individual neurons would be lost. Without being able to distinguish neurons, we could not measure them, test them, work out what they like and do not like. No Nobel Prize for Hubel and Wiesel for their discovery of simple and complex cells in $V1$. No tuning cells in auditory cortex, no place cells in the hippocampus. Perversely, it turns out we have dark neurons to thank for our ability to make sense of the cortex.

THE LONG TAIL

When I say silent, what exactly do I mean? We've been in your brain less than a second. When we make it to a full second, less than 10 percent of your cortical neurons will have fired a spike. In one second, there is

90 percent silence.[11] If we were to hang out here for a whole minute, still the majority of cortical neurons would send no spikes. Yet by tagging along with the spikes from that minority of active neurons, we have reached the farthest ends of the pathways through your visual regions of cortex in a few hundred milliseconds.

Right at the top of this book, I told you there is an average of one spike per second for every neuron in the cortex. But if 90 percent of neurons are silent in one second, how can that be? If so many neurons are silent in a second, then to get an average of one spike per neuron, that must mean some neurons are sending loads of spikes per second. And they are.

About 10 percent of cortical neurons produce half of all spikes. I'll repeat that, because it took me a while to wrap my head around it when I first came across this fact: half of all the spikes in your cortex are sent by just 10 percent of the neurons. In Hromádka's collection of neurons from the first auditory bit of cortex (A1), 16 percent of the neurons contributed half of all recorded spikes. In O'Connor's collection of neurons from the specialized whisker bit of sensory cortex (S1), exactly 10 percent of the neurons contributed half of all recorded spikes. These few are sending the majority of messages, dominating conversation like a myna bird in a monastery. Which means there is a continuum, with truly dark neurons at one extreme, and these myna birds at the other.

To find out exactly what this continuum looked like, in 2012 I surveyed data on the activity of groups of neurons across the cortex, part of my contribution to a mammoth review of neural activity in the cortex written with Adrien Wohrer and Christian Machens.[12] And everywhere I found the same thing. When we choose a chunk of time to watch neurons, the number of spikes per neuron will be unevenly distributed. In that chunk of time some neurons will be silent, most will send just a few spikes, and a handful of neurons will send many spikes. The distribution of activity across a group of cortical neurons is "long-tailed" (figure 6.1).

I found this same long-tailed distribution everywhere. In the first regions of visual and auditory cortex, in motor regions, and in prefrontal cortex; in data using different methods of recording spikes; and regardless

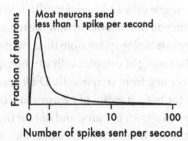

FIGURE 6.1. What I mean by a "long-tailed" distribution of activity. We
imagine recording from a large group of neurons and observing how
many spikes each neuron sends per second. We then work out what
fraction of the neurons send, say, 1 spike per second; or 2 spikes per
second; or 0.1 spikes per second (i.e., a spike every 10 seconds). When
we then plot those fractions, as here, we always see the same thing: a
peak below 1 spike per second, and a long tail way out to the right with a
small fraction of neurons sending 10 spikes or more per second.

of what the animal was doing at the time—quietly sitting, seeing, moving,
or deciding. Always: some silent, most quiet, a few yelling.

My survey of silence has deep implications. For one thing, "average"
activity is useless as a way of understanding what a region of the cortex
is doing. That handful of loud neurons skews our averages upward, a
long way upward, suggesting to us that most neurons are sending spikes,
when they are not. For another, it showed that dark neurons, neurons
sending far less than one spike per second, are everywhere. Most clearly,
it showed that during the brief few seconds of an animal's sitting, seeing,
moving, or thinking, those dark neurons are not communicating anything, to anyone. What then are these dark neurons for?

WHAT DARK NEURONS ARE FOR

Dark neurons are a real pain. Our theories for how bits of the brain work
are based on the patterns of spikes in them. But the dominance of dark
neurons means that our theories of the brain are only about a mere
handful of actual neurons.

Remember the simple cells and complex cells in V1? When we were there, I rehearsed for you the theories of how the tuning of simple cells can be explained by combining spikes from their inputs from the retina, and in turn how the tuning of complex cells can be explained by combining the spikes coming from simple cells. Dark neurons make clear these theories are for but a handful of neurons in V1, those cells that reliably respond when shown pictures, and not for the masses that do not respond. Dark neurons mean that the emergence of brain-like tuning to the visual world in AI networks is not perhaps as insightful as it appears, as this brain-like tuning is a comparison to a relative handful of neurons. And these issues are not restricted to V1. They play out in all areas of cortex, everywhere we have theories of single neurons, across the two visual Highways, across the other senses, across the rest of the cortex.

These dark neurons must do something. Neurons are expensive to build, expensive to maintain, and expensive to use.[13] Your brain uses about 20 percent of all your energy budget, every day. Just keeping your brain cells alive and in good condition uses about 25 percent of the brain's energy budget—that's 5 percent of your total budget, every day. We've already seen that synapses are expensive: about half the moment-to-moment energy used by neurons is in their inputs; the other half is in their spikes.[14] Dark neurons burn energy to stay alive and burn more energy on their inputs yet produce little to no output to show for it. Perhaps one explanation for them is to flip this argument, to think about the energy they aren't using. After all, one way to conserve some of the brain's energy budget is to not send spikes, for without them you halve the energy cost of using a neuron.

But if you didn't need the dark neurons, evolution wouldn't have brought them into existence. And development wouldn't have spent all that energy growing them, dividing them, and extending their axons to the right places. Your body has much better things to spend its energy budget on than dead weight in your head. It would be absurd to fill our visual cortices with neurons that do not see anything. Absurd to grow our massive prefrontal cortex, and then fill it with neurons that sat in darkness. So what are they for? Here are three ideas.

The simplest idea is that in the lab we're not asking the brain to do anything interesting. After all, in the lab we can only probe a tiny fraction of the real-world input these neurons receive. So maybe our experiments are just not rich enough, and we need recordings of neurons in animals behaving naturally over very long times—days, weeks, months—to find out what they are for. Then, if we sample enough of the animal's life, we can find out what the dark neurons respond to. Technically, this is just about within our grasp. Practically, less so. Some poor graduate student has to actually do these recordings, locked in a lab for weeks on end, and lose their entire social life, their romantic partners, and their self-esteem in the process.

As Bruno Olshausen has argued, this dull-world argument is plausible for V1:[15] that the stimuli we use in our experiments are too simple, too poor a reflection of the real world, and all neurons in V1 do respond to something in the world—it's just that we will never find out unless we record them for a whole lifetime. In this idea, the sparsity of activity is also about making the most out of what energy the brain has available. The theory is that V1 has so-called population sparseness, where each neuron in V1 is very selective to what it responds to, so that energy isn't wasted by a swath of neurons responding to the same thing, sending redundant information. And this in turn means there is lifetime sparseness—if those selective things rarely occur, the neurons will rarely send spikes. In this idea dark neurons are then the fault of our limited capacity to probe the brain, not the brain itself.

A second idea is that the dark neurons are a reserve army, waiting to represent new things. The primate life span is long, packed end to end with things to remember, skills to acquire, faces to learn. And for humans, our unique ability to learn new concepts, ideas, and words places heavy demand on our ability to represent things using spikes. Some of that learning will be through changing the effective strength of connections between neurons—at the most extreme, through changing the strength from nothing to something—whether by increasing the size of the voltage blip, increasing the reliability of the synapse, or both. And increasing the effective strength of excitatory inputs to a dark neuron will in turn increase its rate of sending spikes. The upshot is the same as the first idea: the reason we do not know what dark neurons are for is that we track neurons only over the briefest of moments in their entire life span, even in rodents that live but a few years, and so would not see if dark neurons are recruited over time.

This simple idea is complicated by the fact that the brain can't just blithely increase the number of neurons sending spikes. For one thing, this will simply increase the energy drain, so the idea suggests that other neurons would need down-regulating as the reserve neurons are up-regulated. For another, those new spikes need to be balanced—an increase in excitatory spikes needs balancing by an increase in the counteracting inhibition, to prevent runaway explosions of activity.

A third idea is that the dark neurons are sending information just fine. They just do so jointly. Each dark neuron makes a tiny contribution, a single spike every now and again, but since dark neurons make up 90 percent of all neurons, that adds up to a lot of spikes. In this idea, the dark neurons send messages not by many spikes from a few neurons, but by many more spikes from a massive group. And as an individual neuron needs to receive a legion of spikes to make a spike in turn, so this mass of dark neurons could be highly effective. The next chapter will pick up this tale.

What's more, not only are loud shouty neurons in the tiny minority, but also their shouting may well be tuned out by neurons on the receiving end. Remember, synaptic failure can act as a gain control, turning down the influence of shouty neurons, and turning up the influence of quiet ones. Indeed, neurons whose inputs have short-term depression respond best to jumps in which of those inputs are active, not the total rate of inputs.[16] So a group of dark neurons sending sporadic spikes together would be exactly the type of input that such a depression-equipped neuron was looking for. Synaptic failure could, paradoxically, favor neurons that rarely fire.

A nuanced version of this idea also gives us a different view of the lifetime sparseness of neurons. Dark neurons are the vast majority in any region of the cortex. There are likely then many more with the capacity to communicate the same messages than are needed. So perhaps each response evoked from that region, like a picture shown to V1 or a sound played to auditory cortex, draws from a random subset of dark neurons (random from our point of view, not the brain's). That random subset sends a spike or two, then shuts down. So, from our point of view, most neurons are silent most of the time. Yet the message sent by the dark neuron population is the same each time.

This random recruitment idea has a simple experimental prediction. Record from a group of neurons while repeating the same event over and over again—showing the same picture to V1, or making the same movement of an arm. Then most neurons should vary whether they respond or not on each repeat of the event, apparently at random. We see just this random participation in groups of whisker-sensing neurons across hits on their whisker;[17] groups of arm movement neurons across similar movements;[18] and even in groups of crawling neurons across repeated bouts of crawling in sea slugs.[19] Dark neurons are then not dark, just misunderstood.

Three ideas for what dark neurons are for, three ideas for solving the paradox of how neurons that send no spikes somehow contribute to the life of the brain. By contrast, those minority of neurons that reliably send spikes, those in the long tail, they should be easier to understand, right? No. Many of those seem to be talking without listening.

TALKING WITHOUT LISTENING

Hidden in the active neurons are many of another kind of dark neuron, a Type 2 dark neuron. Neurons firing away just fine, spikes streaming out. But that do not seem to respond to anything, whose output of spikes does not meaningfully change no matter what is happening in the outside world. Talking to other neurons, but apparently not listening. Dark to the outside world.

These Type 2 dark neurons were hiding in plain sight in huge piles of neuron recording papers from the 1960s through to the early 2000s. In those papers, the first sentence out of the blocks in the Results was always something like: "we recorded a total of N neurons, one at a time, while presenting stimulus X or invoking movement Y. We found M out of the N neurons responded to the presentation and will be subject of this paper." Always, the number M of responsive neurons was far less than the total number of recorded neurons N, and the rest were just thrown away! So what were those other neurons, the N-M unresponsive neurons, doing?

Type 2 dark neurons are especially clear in modern recordings of many neurons at the same time. For with these recordings we can take huge samples of neurons, numbering in the hundreds or thousands, and examine each neuron individually for its tuning, for what elements of the world it responds to. When Simon Peron and colleagues looked at their recordings from that specialized bit of whisker cortex, they dubbed 67 percent of those neurons as "active"—and that included neurons firing barely at all. Searching for what the spikes from those 67 percent were responding to, they found 28 percent of them were sending spikes with no detectable task information at all. No tuning to whether the whisker was moving, or hitting a pole; no apparent difference in whether the pole was in one position or another, or the trial was to lick left or lick right. Twenty-eight percent of the active neurons in the very first bit of cortex getting input from the one and only whisker being used were not apparently listening to what that whisker was saying.

Christopher Harvey and colleagues found a similar story in their recordings from the parietal cortex of those mice running in a virtual reality maze. Remember, they had dubbed just 47 percent of their neurons as "active." And of those, 27 percent were sending spikes with no task information at all. No tuning to where the mouse was, or what it was doing. Active neurons in parietal cortex seemingly did not listen to what was happening to the world around the mouse.

My lab found the same thing in the prefrontal cortex, in recordings of neurons that came from rats running up and down a Y-shaped maze.[20] In each session of training on the maze, our collaborators had recorded between 12 and 55 neurons. And all were active, or at least not dark, as these were electrode recordings. Yet in each session typically just one or two of the neurons responded differently to the things happening in the world around the rat. Only one or two sent different amounts of spikes whether the rat chose the left or the right arm; whether the rat got a reward or not at the end of the arm it chose; whether or not the

light was on at the end of the arm. It was like the active neurons in prefrontal cortex just didn't care about what was going on (but they very
much did, and how they did is the story of the next chapter).

You may be wondering: what do you mean exactly by "respond"?
That innocuous question opens up a deep hole. Typically, it means that
a neuron has sent a different number of spikes in one condition compared to another. Like whether a whisker hit a pole or not. Or whether
the animal has gone left or gone right. And "different" is defined by
setting a threshold, typically by asking whether the difference in the
number of spikes between the two conditions passes some statistical
test. This is how we end up with the ubiquitous idea that there are different functional types of neuron in the cortex. That neurons all over the
cortex are either the type that respond to thing X, or they do not. Neurons in V1 are either simple cells or they are not. Neurons in auditory
cortex are either tuned to a specific group of frequencies, or they are not.
Neurons in somatosensory cortex are either tuned to the touch of a finger on a surface, or they are not. Neurons in motor bits of the cortex are
either tuned to the speed of arm movement, or they are not. Neurons in
prefrontal cortex either respond to the value of a reward, or they do not.

But as Adrien Wohrer, Christian Machens, and I pointed out in
2013,[21] using a dividing line to label neurons as either "responding" or
"not responding" is a fallacy. It breaks into two groups the continuum
of responses from the nonexistent, through the weak, the moderate, the
middling, to the exuberant. We can always move the line and break the
continuum into two different groups. And this means there are not
really "types" of neurons we can define by their responses to the world.

We can reel off many examples where apparent neuron types do not
exist; where neurons sit on a continuum for how they change their
stream of spikes. We've already seen for V1 that simple and complex-like
cells sit on a continuum. In the prefrontal cortex of monkeys touching
a vibrating bit of metal, there is a continuum for how much the neurons
there respond to the vibration's frequency.[22] In the parietal cortex of rats
making left-or-right decisions based on the number of clicks or flashes,
both the neurons' preferences for direction and their preference between clicks and flashes lie on a continuum.[23]

Which all means Type 2 darkness is one end of a continuum of responsiveness. What we've been calling Type 2 neurons are those that barely change their spiking to an event in the outside world, a change too weak to be picked up by our analyses. But are the changes too weak to be picked up by other neurons in the brain? After all, what if these weak changes in the stream of spikes happened in many neurons at the same time? To answer this properly, it's time to turn to the deepest question in neuroscience: the meaning of spike.

CHAPTER 7

The Meaning of Spike

THE COUNTERS AND THE TIMERS

For nearly a century, a war over the meaning of spike has been fought between the Counters and the Timers.[1]

The Counters believe a neuron sends its messages in the number of spikes it emits. They think spikes carry meaning by their number. The Timers believe a neuron sends its messages by when it emits a spike. They think spikes carry meaning by when they occur, especially in relation to each other.

The war has taxed our best minds since the first spikes were captured by Lord Edgar Adrian, Joseph Erlanger, and others in the 1920s. The evidence marshaled on both sides is by now formidable.

The Counters

The Counters are dominant. Endless studies have asked neurons about their preferred thing in the world, about what makes them send the most spikes. Because doing so is simple. Present some sensory thing—a tone, a surface, a line. Then vary it. Change the frequency of the tone, the roughness of the surface, the angle of the line. And simply count the number of spikes the neuron sends as you vary the sensory thing. Voilà, a tuning curve (figure 7.1). You now know what frequency, or roughness, or angle makes your neuron send the most spikes. We could then claim that spikes from this neuron mean a tone of a particular frequency, a

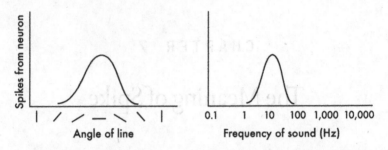

FIGURE 7.1. Tuning curves—how Counters see neuron coding. These are cartoons of two different types of tuning curves. On the left, tuning curves of a hypothetical neuron in V1, which we have presented with lines at different angles and then plotted the number of spikes it sends in response to each angle: this neuron prefers lines that are horizontal or close to it, but detests vertical lines—loves the horizon, is oblivious to skyscrapers. On the right, the tuning curve of a hypothetical neuron in the first auditory area of cortex (A1), which we have presented with different frequencies of sound and then plotted the number of spikes it sends in response to each frequency: this neuron prefers sounds about 20 Hz, but does not respond above about 200 Hz—adores teeth-rattling sub-bass, unmoved by arias.

surface of a particular roughness, an edge at a certain angle—a simple cell. For Counters, meaning is simple.

We can do the same trick for movements. With the subtle but important twist that now we're not counting spikes in response to something, after something, but counting spikes that happened just before an event. Record a neuron while an animal is moving its arm repeatedly at different angles. Then simply count the spikes that happened just before the start of each arm movement. Result: a tuning curve for what angle of movement that neuron prefers.[2] The same trick works for speeds of the same movement, or for contractions of individual muscles, or more complex combinations.[3] We can infer the counting code by working backward from what the animal is doing to what the neuron was doing just beforehand.

This reverse inference of tuning works for more complex properties of the world too. A famous Nobel-prize-winning example is the coding of place.[4] Watch an animal run around a big box or maze, all the while recording from a neuron in its hippocampus. Counting the spikes from that neuron will reveal that it has a preferred location, that it sends the most spikes in a particular place, fewer spikes when close to that place, and no spikes when far from it. It is a place cell.

And if we scour the regions surrounding the hippocampus, we find a menagerie of counting neurons.[5] Poke around in there and you'll find the head direction cell, whose maximum count of spikes will tell you the particular direction the animal is facing; the boundary cell, whose maximum count of spikes will tell you the animal is at or close to a boundary in a particular direction (e.g., to the east); and the grid cell, whose maximum count of spikes reoccurs periodically in space, as though it has laid a grid over the world and then sends the most spikes every time the animal reaches an intersection of that grid. All single neurons whose count of spikes indicates some property of physical locations in the world.

Neurons that send messages in their number of spikes are everywhere, it seems. But perhaps we only see a counting code reported everywhere because counting spikes is easy for us. Because the default reflex of the experimenter is to count spikes and then report the number of spikes, rather than test more complex ideas of how neurons send messages. After all, when Timers do make more nuanced measurements, they can marshal some extraordinary evidence too.

The Timers

We've already met some of it. Indeed, I already told you in chapter 2 that one of the deep reasons for the existence of spikes is accuracy—the sending of information at precise times. We learned there that repeating the same movement of a rat's whisker causes the first neuron in the whisker system to send the same pattern of spikes with ridiculous submillisecond precision. Other sensory systems show similarly accurate timing of spikes.

The owl's hearing system is home to the most complete circuit for spike timing yet worked out.[6] Small woodland rodents know the remarkable capabilities of this circuit all too well. Owls can accurately determine the location of a scurrying rodent from sound alone. Their brain does this by using the difference in timing between a sound

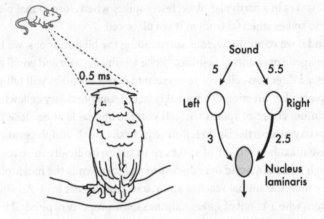

FIGURE 7.2. Precise timing of spikes in the owl's hearing circuit. The sound
of a mouse 30 degrees to the left of the owl's head will arrive slightly earlier
at the left ear than the right ear (here, 0.5 milliseconds earlier). Now
consider the circuit of neurons on the right; numbers indicate the speed of
transmission in milliseconds. The first neurons receiving sound from the left
ear will respond half a millisecond earlier than the neurons receiving sound
from the right ear. Output from the left- and right-side neurons converge
on neurons in the nucleus laminaris. There is a particular set of right-side
neurons whose axons are just thicker enough that they transmit half a
millisecond faster than the left-side neurons. Which means this particular
nucleus laminaris neuron will receive spikes from the left- and right-side
neurons at the same time—and send its own spike, now meaning
"the sound is 30 degrees to the left."

arriving at the left and right ear. If the noise is straight ahead, the sound
arrives at both ears at the same time. If the noise is from the left the
sound arrives at the left ear a few milliseconds at most before the right
ear; and vice versa for sounds from the right. And the exact delay between the sound arriving at the two ears is proportional to the angle
between the sound and the owl's head—the farther to the left, the longer the gap between arriving at the left and right ear. Longer, but we're
talking differences of less than a millisecond here.

It turns out each axon from the first group of neurons in the owl's
hearing system has a very specific, very precise delay in sending a spike
to a sound arriving at their ear (figure 7.2). And these first sets of axons

from the left and right ear converge on a second group of neurons, the nucleus laminaris. A neuron here will send a spike if the spikes from the left-ear neurons and the right-ear neurons arrive at the same time. But if there is a delay between when a sound arrives at each ear, how can the spikes arrive at the same time? This is where those very precise delays come in.

Let's say a neuron in this second group is a detector for sounds at 30 degrees to the left. This means it should fire spikes if a sound arrives at the left ear before the right with the specific delay corresponding to 30 degrees. To do this it will receive input from a specific set of axons from the left ear and a specific set of *faster* axons from the right ear. Crucially, the right ear axons are exactly faster than the left ear axons to cancel the delay between the sound arriving earlier at the left ear and later at the right ear. Cancels to within less than a millisecond. Owls catch mice using a hyperprecise spike timing code for the locations of sounds.

Precisely timed spikes to sounds is not restricted to owls. In rodents, the first bit of cortex that receives input from the ears produces what DeWeese and colleagues called "binary spiking."[7] When played a sound, a neuron here either sends a single spike at the onset of the sound or does nothing. And if a sound does elicit a single spike, repeating it will elicit each time a spike at the same delay from the start of the sound. Hence binary: a spike (1), or not (0), at a precise time after the sound starts.

The output neurons of the retina—the ganglion cells—show similarly precise delays in spikes. Each time a ganglion cell is presented a pattern of flickering pixels (a pattern in its part of visual space) the first spike it sends is at the same delay to within a few milliseconds.[8] Building on this, Tim Gollisch and Markus Meister showed in 2008 that each ganglion cell seems to have a spike-latency code—for each different image it is shown, a cell sends its first spike at a different latency, but repeat the same image, and the same latency occurs.[9] The latency of the first spike encodes far more information (literally, in bits) than counting spikes about which image was presented. Across a big group of ganglion cells, Gollisch and Meister could use their spike latency alone to reconstruct the presented picture. A fuzzy, gray-scale picture, admittedly. But still, the timing of spikes in the retina seems to be a powerful code.

Different Strokes?

You, being smart, may have noticed something that decades of neuro-scientists had not: the Counters look at one set of brain regions, and the Timers look at another. Often they look at very different species. To those looking deep within the cortex, or the hippocampus, or the amygdala, or at motor neurons in the spine and brain stem, the counts of spikes make the most sense. To those looking at the first steps of the sensory systems, in the retina, at the first brain regions getting inputs from the ear, or the whiskers, timing and patterns are everywhere.

So is that the answer? Brain regions at the edge, especially those using sensory information, use timing; brain regions in the middle, especially the cortex, use counting?

Strong evidence for this would be that cortical neurons cannot even send precisely timed spikes. After all, we already saw in chapter 3 that they send spikes with irregular spacing, almost perfectly capturing a random process. How can such spikes carry timing information if they are occurring "randomly"?

The simplest test is to give a cortical neuron an identical input many times and see if it repeats the same pattern of spikes with the same precision (figure 7.3). An input direct to its body, bypassing the unreliability of synaptic failure, and the crushing inhibition by GABA, to remove them from the equation. We know what happens. Inject a pyramidal neuron's body with a constant blast of current, and it reels off a set of spikes. Give it the same constant blast of current again, and the set of spikes do not repeat exactly.[10] If cortical neurons cannot even pass this simple test, surely they cannot be sending messages based on the timing of spikes.

But instead inject a noisy, random current, something mimicking the wild fluctuations of the neuron's voltage at its body as it's bombarded by the flickers of voltage cascading down its dendrites. If we repeatedly inject the same noisy current, we get precisely the same response, the exact same random-looking pattern of spikes, each time! Neurons in the cortex are capable of reproducing exactly the same timing pattern of spikes given the same noisy input. And there are hints that they do use a timing code.

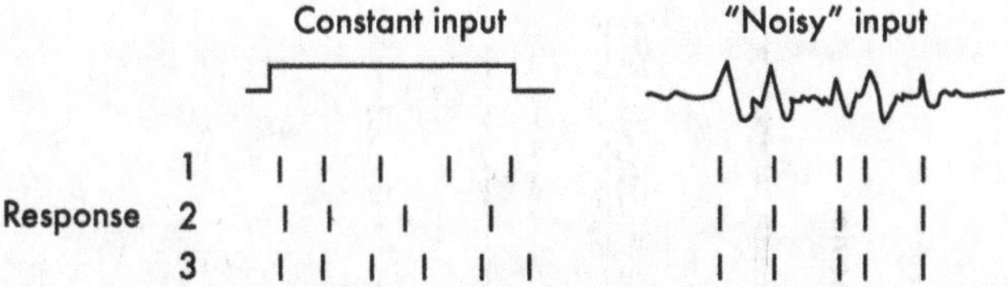

FIGURE 7.3. Precise spikes from a cortical neuron depend on its input. We imagine injecting a current directly into the body of a cortical neuron and repeating this three times to see how similar its response is each time. For a constant input (*left*), the neuron will spike as soon as the input turns on, but thereafter the timing of the spikes starts to differ between the three repeats; indeed, the neuron sends five spikes on the first go, four on the second, and six on the third. Identical input, to the same neuron, but different times of sending spikes. But if we instead inject the same noisy input each time (*right*), then the neuron sends the spikes at the same time on each of the three repeats.

A big hint comes from area MT. We left area MT a few synapses ago, as its neurons sent volleys of spikes responding to the main directions of motion in the world. To the coherent collection of edges and angles you call "Sarah" striding across the office, as she moves across your field of view from left to right. How we know that area MT sends spikes to motion is that we asked some monkeys to watch some moving dots.

These bored monkeys watch movies of randomly moving dots, trying to decide which direction the dots are moving in. Sometimes the dots are coherent, mostly moving in the same direction, so the motion is easy to see. Sometimes the dots are really noisy, with only a few moving together in same direction, so the motion is hard to see. Sometimes in between; and sometimes impossible, when all the dots are moving at random. Watching these movies evokes spikes from area MT neurons. Ones that like leftward movement spike when dots move left; those that like movement along 30 degrees from upright spike when dots move in that direction. Like most cortical neurons, area MT neurons send these spikes seemingly at random, the spacing between them highly irregular—some clumped, some far apart. Yet Wyeth Bair and Christof Koch revealed in 1996 that if we repeat the exact same movie of random dot motion then we see the exact same sequence of apparently random spikes from an area MT neuron[11] (figure 7.4).

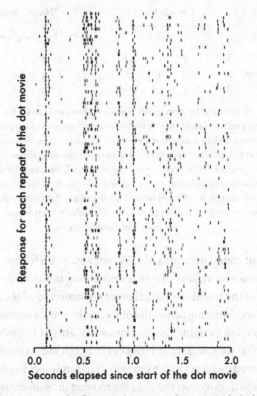

FIGURE 7.4. Precise spikes from a single neuron in the cortex. We're looking at the spikes sent by a single neuron in area MT, while the monkey it belongs to is watching many repeats of the exact same movie of dots moving randomly. Each tick is a spike from that neuron; each row of ticks the spikes it sent during one presentation of that same two-second-long movie of dots randomly moving about. A single row looks random—there are small and big intervals between the spikes within a row. But they line up across the rows: if we start at time 0, when the movie kicks in, we can see the ticks repeatedly line up top to bottom at the same times in the movie (e.g., at about 0.1 seconds, the cluster just after 0.5 seconds, and again at about 1 second). This neuron sends its spikes at the same points in the movie each time it is shown. (Redrawn from Bair and Koch, *Neural Computation* 8 [1996]: 1185–201.)

As we departed on the cloned spike from the area MT neuron, we were part of that neuron's volley of randomly spaced spikes, part of its response to all things moving in the office. The dot motion results show that if the exact same set of movements across the office repeated exactly the same way—Sarah striding, Graham nonchalantly fiddling with his tie—then we'd have been on the exact same spike at the exact same time. A timing code in cortex. But a hidden code: it would never show up in the real world, as the world never repeats itself so exactly.

Yet we can also construct seemingly cast-iron arguments that a cortical neuron cannot be using such timing codes. A compelling one is that spikes can be knocked off course too easily to use a timing code.[12]

This is easy to demonstrate. Build a computer model of a bit of cortex: create thousands of artificial neurons that send spikes, wire them together, give them all some input, and watch. They will each produce a characteristic and random-looking pattern of spikes. Repeat with the exact same input, and the exact same patterns will result. But now repeat it again, and this time delete a single spike sent by one neuron. As you watch, the spikes of many of the other neurons will quickly wander off into new patterns, some radically different from before.[13] Clearly, such neurons could not be using a timing code if a single failed spike can change the course of so many other neurons' spikes. And we know spikes fail all the time.

How do we bridge these irreconcilable differences between Counters and Timers? As we cling to our spike racing into the prefrontal cortex, what can we say about what it means? We can look back to where we came from.

PREDICT ME

Aiming for a fragile peace, others have sought to directly decode the message of spikes by finding out what predicts them, or what they predict.

At the beginning of our journey, prediction seemed simple. Coming from the first bits of visual cortex, those just a leap or two from the eye, we traversed spikes evoked by a particular type of edge or color or direction of movement, in a particular pixel of the visible world. Flipping this around, the presence of a particular edge, color, or movement in a particular pixel thus predicts a spike from a particular set of neurons in V1. Likewise, a spike streaming from the first bits of auditory cortex is evoked by something about the basic properties of sound, its frequency or volume or direction. The occurrence of a particular frequency at a sufficient volume in the world will predict there should be a spike from a particular set of neurons. And so we could say: if thing X predicts a spike, this is what the spike means.

Yet even here, in these bits of cortex closest to the machinery for sensing the outside world, we cannot predict every spike a neuron sends from a simple line or a simple sound. A neuron sends many spikes that are not obviously evoked by a particular thing at a particular time. So what do those spikes mean? Some smart people hit on a solution: ask the spikes themselves.

The basic idea is simple. For every spike a neuron sends, find out what was happening in the world just before it. And the clever part is that we don't guess. We learn what was happening directly from the data.

The goal is to create a model that takes as input what happened in the world in the past few hundred milliseconds or so and outputs a prediction of how likely a spike is to occur right now. In other versions, such a model might instead predict the likely number of spikes to occur right about now. Such models can either be a Timer or a Counter, depending on how long we define as "now": if the model is predicting a few milliseconds at a time, we're building a timer; a few hundred milliseconds at a time, we're building a counter. Indeed, building models to predict spikes drives home that the dividing line between "timers" and "counters" is a fuzzy one.

To make a spike prediction, we give the model measurements of things in the world—like the angles in a picture, or the frequencies of a sound—over those few hundred milliseconds in the past. And the model assigns a weight to each of these measurements at each moment in the past. The higher the weight, the more influence that measurement at that particular time in the past has on the probability of a spike right now. The model's prediction comes from adding up all the different measurements, over each of those moments in (a very brief) history, and saying: right now I predict a spike is imminent (or not).

The key is that the model learns the weights. It changes them until its predictions match the real spikes as closely as possible. Then once it's done learning, we seek out the highest weights and voilà: we find out what aspects of the world at what times in the past the neuron sends spikes to![14]

This predictive model approach works best in the early parts of the brain's sensory systems, those right up close to the sensory machinery. In the retina, these predictive models reveal precisely how the exact location and timing of changes in the visual world predict when retinal ganglion cells will send spikes.[15] Better, the predictive models fall into two types of cell, one predicting spikes after a sudden increase in light level, the other after its sudden decrease. These are precisely the ON and OFF type retinal ganglion cells we met in chapter 2, now with their existence confirmed directly by learning from the data.

These predictive models are also highly effective in predicting spikes from those first neurons to get input from the whiskers, those ultraprecise timers of spikes. There, the models learned that sudden changes in the bend of a whisker accurately predict spikes from these neurons.[16] And what causes a sudden bend of a whisker? Hitting something. So these neurons send very precise signals about when and how a whisker strikes an object in the world. Equally important, the models learned that the angle of the whisker does not predict spikes at all. By predicting spikes, these models tell us what the whiskers can and cannot tell the rest of the rodent brain about the world.

But the deeper we have plunged into the cortex, the less we can predict from a spike about what is going on in the outside world.[17] Predictive models can tell us little about the meaning of spikes deeper in the brain, about the spikes we've clung to for the past few jumps. For they fail to predict more than a few percent of the spikes from a neuron. They struggle even in the first whisker bit of cortex, just three jumps across synaptic gaps from those primary whisker neurons whose spikes are so well predicted by the bend of the whiskers.[18]

We already know one key reason why prediction methods may fail: dendrites. As we learned in chapter 3, the barrage of inputs coming into a neuron's tree are not just weighted and summed up, but can be dramatically transformed. And that transformation can radically change the relationship between the outside world and the spikes sent by a neuron. Indeed, when attempting to get meaningful predictions of spikes in that bit of cortex dedicated to whiskers, Peron and friends had to build a model that learned how to transform the events in the worlds, the whisker bends and angles, into a complex, messy form just to get anything close to predicting the output of a single neuron. But as we get deeper into the cortex, having jumped many gaps onto many neurons, the biggest barrier isn't dendrites; the biggest barrier to inferring meaning from prediction is, simply, other neurons.

You may be wondering: where are we, exactly? And that's a great question. The vagueness is deliberate. We traveled along Highway What, and by cloning ourselves traveled along Highway Do at the same time. Highway What has disgorged us into the start of the enigmatic prefrontal region of the cortex, the front third of your cortex, roughly everything forward of your ears. Highway Do left us in part of the parietal regions of cortex, a large strip above and behind your ears (figure 7.5). It becomes harder now to know precisely what any particular small bit of these regions of cortex do. And their roles are intertwined. Neurons in both prefrontal and parietal regions send axons to the other, so we could hop a spike and jump between them if we so wished.

Indeed, the prefrontal and parietal cortices are the brain's Louvre: vastly too much to take in on one visit, and no matter how niche your tastes you can find something to suit. The prelimbic region, to pick one of many, is like the ceramics gallery: if that's your bag, great—you could spend all day in here marveling at how its neurons send spikes after a mistake is made, seemingly in order to ensure you take more time before making the same decision again;[19] if it's not your bag, glance at the serried ranks of plates as you hurry to the exit. If we looked hard enough, I'm pretty sure we'd find a neuron in the prefrontal cortex that only fires when a rat turns left on the second Wednesday after Pentecost while wearing a fez. Because there are simply so damn many of them, we can always find a neuron here that seems to correlate with something.

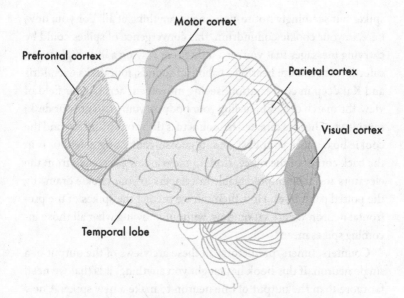

FIGURE 7.5. Key regions of the cortex on our journey.

By my reckoning, we've only a few tens of milliseconds at best between our arrival in these cortices and our exit to the motor cortex. And that's if we take a languid path. So as we leap from axon to dendrite across microscopic gaps, follow the cascade of neurons up and down the layers, plunge through the white matter to jump between regions, I will point you to the main attractions, the Venus de Milos, the Mona Lisas, the Marly Horses. As we grab a cloned spike charging deep into the prefrontal cortex, the first of these is the loneliest neuron.

Farthest from the inputs from your senses, farthest from the outputs to your muscles. It will never know the taste of pizza, the smell of fresh bread, the deep red of a sunset, the touch of a baby's hand. But it will receive distant echoes of all these things. Our spike and millions like it arrive bringing their messages from all over the cortex to the loneliest neuron.

And that's the key. If we only look at what happens in the outside world, then we discount the influence of other neurons on that neuron—other neurons carrying all sorts of information, information that we cannot guess or, worse, those Type 2 dark neurons, sending spikes but seemingly not responding to anything at all. For you, now, here, in your cookie conundrum, this convergence of spikes could be carrying messages that you're: a little tired; craving a little high-sugar, calorific pick-me-up; hearing a babble of sounds, four desks over, Idris and Kai deep in conversation; seeing movement across your field of view, the march of Sarah; feeling your body on your chair at your desk; sensing your head is directed to look across the room at and beyond the cookie box; aware your desk and its cookie-containing neighbor is in the back corner of the office, close to the windows, screened from the elevators and their unpredictable interlopers in your cookie drama by the potted palm trees. How then can we predict the spikes of the prefrontal neuron we are sat on now, without knowing what all those incoming spikes mean?

Counters, timers, predictors: all these are views of the output of a single neuron. If this book has taught you anything, it is that we need far more than the output of one neuron to make a new spike. A new spike is a summary of the total messages from the input of hundreds of neurons in both space (who sent a spike) and time (when they sent a spike). Counters and Timers are asking the wrong question. Meaning is not the spikes of one neuron, but of the legion. Ask not what a neuron sends, but what it receives.

WHO SENDS SPIKES

Asking the legion's meaning once seemed the realm of science fiction.[20] But the current golden age of systems neuroscience has changed all that. The recent explosive advances in recording technology mean now we can at last record hundreds of neurons at the same time. Rather than try to predict the world from the spikes of a single neuron, recording many neurons gives us a different way of thinking about the meaning of spike. Not how many or when, but who: which neurons send spikes at the same time.

So now my lab and others ask instead, can we predict what is happening in the outside world from the pattern of neurons who are sending spikes? Imagine recording from three neurons in a bit of the cortex that cares about vision, while we repeatedly show the eyes one of two pictures—one of the sought-after pear, ginger, and chocolate cookie, and one of a green cuddly toy dragon (named Steve). If these three neurons together encoded something about the difference between these two pictures, then they should send between them a markedly different pattern of spikes to the cookie picture than to the dragon picture. But they need not always send the exact same pattern to the same picture. Indeed, we've already seen that a single neuron's response to the same event in the world can sometimes be highly repeatable, yet most of the time for most neurons it is quite variable, to the point of occasionally not responding at all. But by looking at more than one neuron, we can find whether the pattern of neurons sending spikes is broadly similar, even if some individual neurons are making a mess of things.

How we find these broadly similar patterns is, again, by learning them. We ask if we can predict the picture from the pattern of activity. We take some of the patterns during the cookie picture and during the dragon picture and train a model to find the consistent differences between the patterns. Then we give the model more patterns of spikes across those same three neurons, evoked by the same pictures, and ask if we can predict which picture was being shown. This is population decoding: take the activity of the recorded legion—ten, twenty, a hundred neurons—then see if we can tell the difference between the legion's pattern of activity after thing X (a picture, a sound) or before thing Y (a choice, a movement).[21] All over the cortex, the answer is yes: we can near-perfectly predict what was happening in the outside world from the pattern of active neurons.

Back in V1, if we'd gathered spikes from the neurons around us as we left the first simple cell, we could have perfectly decoded the angle of the line passing through the pixels in the world our neuron and its neighbors were looking at.[22] We would have seen a unique pattern of which neurons were sending how many spikes, all in response to that beautiful slant of the cookie's crumbly top. Just ten or so neurons would have been enough to detect the pattern, as it emerged just a few tens of milliseconds after your eye fell on the cookie.

Better still, had we known about those Type 2 dark neurons, those that talk without listening, we could have wielded our population decoder in V1 to show us they really do something. For example, Joel Zylberberg took some recordings of hundreds of neurons from the visual cortices of mice that were staring at lines moving in one of eight different directions.[23] Among those hundreds of neurons, some had clear preferences for one of those directions of movement; many had no preference at all—they were active but not listening, yet more awkward Type 2 dark neurons. Zylberberg found he could decode the direction of movement consistently better from a mixture of tuned and dark neurons than from tuned neurons alone.

It's a similar story for whiskers. In that special bit of rodent cortex dedicated to whiskers, researchers in Miguel Maravall's lab at the University of Sussex found that they could not use the spikes sent by a small group of neurons to work out whether the whiskers were rubbing a rough or smooth texture.[24] These small groups, perhaps at most four or five neurons, were seemingly not listening to the whiskers and the messages they were sending about their jolly time rubbing against sandpapers of different roughness. But combine even just three such dark-neuron groups, and the difference between rough and smooth is readily detectable. Thus population decoding can show us the spikes of a population of neurons carry meaning that's seemingly invisible when we look at members of that population.

Now we're in the prefrontal cortex, what can we learn by wielding our population decoder, by looking at the pattern of spikes sent by our neuron and its neighbors? We learn that, even in simple tasks, we can decode many complex things. We can indeed decode messages arriving from all over brain.

Wield the decoder in the prefrontal cortex of monkeys staring at screens, and we can decode a lot from the legion. Ask a monkey to watch seemingly endless sequences of pairs of pictures, and from the pattern of spikes at different points in the task, we could perfectly predict which of four pictures was presented first; which of a different four was presented second; even exactly what the monkey was supposed to do with that sequence—commit it to memory, compare it to memory, or move its eyes between them.[25] Ask a monkey to look at a four-by-four grid of LEDs and see which one lights up; lo and behold, we could decode from the legion of neurons in prefrontal cortex which of the sixteen LEDS is lit.[26]

We can even decode different features of the world from the same legion at the same time. Ask a rat to run up a Y-shaped maze, to make a choice about whether to go down the left or right arm of the "Y," and from the patterns of spikes in that rat's prefrontal cortex, we could decode (indeed, Silvia Maggi in my lab did decode) the choice the rat was about to make; separately, we could decode which arm of the maze was lit at the end.[27] The same small population of neurons knew about things in the world right now—the lights at the far end of the arms, visible to the rat's eyes—and about internal things—the choice of movement that was about to happen. Quite the convergence of messages.

At the end of Highway Do we'd have the same ability to decode multiple features of the world from the spikes of the legion of neurons around us. David Raposo and Matthew Kaufman in Anne Churchland's lab asked their rats to master a complex task of counting clicks or flashes of light (or both), and using that count to decide whether their reward would be in the left or right food dispenser—left for a low count, right for a high count.[28] In the back end of parietal cortex, from the legion of neurons they could decode whether the rat was being played the clicks or flashed the lights, and separately they could decode the upcoming choice of going left or right. In both the work of the Churchland lab and our own, the patterns of spikes from the legion of neurons meant different things at the same time, depending on how we read them out.

(Before we get carried away with our apparently deep understanding of the brain, a word about the decoding fallacy. Just because we can decode information about thing X from spikes, does this mean the brain actually has access to this information?[29] Say we can decode from the activity of a hundred neurons whether a light is on or off. Does this mean the brain "knows" the light is on or off? Not necessarily. It definitely means there is something different between these states of the world, and we can decode from the brain that they are different. But something else in the world could always be different at the same time as the thing we're interested in, something we haven't noticed—like the light switch being in a different position when the light is on or off—and that's what the brain actually knows about. But we can check if the brain seems to know about what we decode, by showing that the thing we can decode, or how we decode it, has consequences: that it is related to behavior, or predicts other neural activity.)

The loneliest neuron is indeed receiving converging messages from all over the cortex. From the legion of neurons in prefrontal and parietal cortices, at the ends of Highway What and Highway Do, we can decode many things about the world, indeed many different features of the world at the same time. And those converging messages are crucial for two things, things vital for solving your cookie conundrum. For the spikes of the loneliest neurons do not just mean what's happening now. They also send messages about the past and the future. About holding the world in memory, and making a decision.

HOLD THAT THOUGHT

Gathering reams of information about the world—the cookie, the box, the desk, the people, their movements—is useless if you can't hold it in mind. Without some form of short-term buffer, some snapshot of all the things in the world you've already noticed, every moment you'd have to re-look or re-hear or re-read everything in order to know what is going on in the world. Like knowing that Adam is not at his desk behind you, as he popped out a minute ago for some "fresh air" while suspiciously pocketing a small rectangular box, so he can't interfere with the cookie conundrum. Or not being constantly surprised that you're sitting in a chair.

If you wanted such a buffer in your brain, you'd need two things. You'd need the messages of what is happening in the moment to converge on a single location. The sights, the sounds, the places, the people, the faces—the messages of them all converging to create the snapshot of the world, now. And you'd need the neurons receiving those messages to buffer them, to not just send one or two spikes and go silent, but send a stream of spikes for as long as you'll need to hold that snapshot in mind. And here in the prefrontal cortex is the perfect place for that buffer.

We've long known that some regions of the prefrontal cortex must act as a memory buffer. Damage chunks of the prefrontal cortex and this short-term memory breaks.[30] Such damage stops you from holding an item in mind for more than a few milliseconds. Say you'd damaged your prefrontal cortex. If I showed you two boxes, one holding a cookie, closed and covered them for a few seconds, then uncovered them and asked you to point to the box with the cookie, you'd have no idea which one it was. Which means prefrontal cortex should have neurons that sustain spiking throughout a buffered memory.

As we leap through regions of the prefrontal cortex, we jump gaps with cloned spikes onto neurons that sure look like they're holding memories. As we arrive in the body of one such neuron, part of the voltage blip entourage descending its tree, we can see from the jumbled mass of ions inside that it sent a spike just a few tens of milliseconds ago, one a few tens before that; we can feel our blip and its entourage have set off the runaway cascade to create a new spike, and with a wave of voltage blips on the way down the tree behind us the creation of more spikes is imminent. The spikes are sustained, but is this a memory?

We can find out by setting the brain a working-memory task. Solving such tasks is only possible by holding a piece of information in a memory buffer during a delay. Like remembering which food hopper had the treat inside before it was covered up. Or a rat sitting in the center of a cross-shaped maze, holding in mind which arm of the maze it went down last time, because it's not allowed to go down the same arm twice. In these working-memory tasks single neurons in prefrontal cortex reliably send spikes throughout the delay period, as though sustaining the memory.[31] We can even show that the memories are specific, that single neurons are sending spikes to different aspects of the information that needs remembering.

In a classic example from 1989, Funahashi and colleagues from Patricia Goldman-Rakic's lab showed that when monkeys were remembering which of a ring of eight lightbulbs had flashed seconds earlier, some neurons in their prefrontal cortex sent many spikes, as though sustaining the memory of the flashed bulb.[32] Most such neurons were selective for a particular location: each of these sent the most spikes to a flashed bulb at a particular point in the ring—and the farther the actual flashed bulb was from this preferred location, the fewer spikes it sent. Each buffering neuron was holding a specific memory of which bulb had flashed. Similarly, a string of studies from researchers in the lab of Ranulfo Romo have shown that neurons in prefrontal cortex send spikes while a monkey is remembering how rapidly a metal strip had vibrated against its fingertip.[33] Here too some buffering neurons are selective, sending spikes in proportion to the speed of the vibration. These complex memories of where bulbs are flashed and how fast a thing is vibrating are created using different senses yet held in the same place, the prefrontal cortex, thanks to the convergence of spikes onto the loneliest neurons.

Again though, Type 2 dark neurons plague the prefrontal cortex. Most neurons here, even those sending spikes, pay apparently little attention to the need to remember things. Funahashi and colleagues recorded 288 neurons in total, yet only 87 of them, just 30 percent, showed a consistent change in their spiking in the delay between the flashed light and the signal to go. Yet if we look at a group of neurons in prefrontal cortex, a sustained memory is crystal clear.[34]

Their combined activity contains a perfect memory of how fast that metal strip had vibrated against the fingertip.[35] Even a memory of how much time has elapsed since the strip had vibrated. Using population decoding, we can see the memory of exactly which one of sixteen different lights was briefly flashed.[36] Better still, we can see that groups of neurons in prefrontal cortex maintain a memory even when we don't explicitly tell it to.

My team and I went looking in the prefrontal cortex of rats trudging back along a Y-maze, having just made their choice of arm—left or right—and having just discovered if they had chosen correctly or not.[37] If they had, then glorious chocolate milk was their reward. We wanted to know what the rats were mulling over as they trudged back to the start—were they reflecting on their failure, or reveling in their triumph? Remembering that going left had been good, or that going right was a waste of time? Every time we looked, just a handful of single neurons, at most 20 percent and sometimes none at all, were firing differently after a left or right choice, or differently after getting chocolate milk or not. The vast majority were Type 2 dark neurons, spiking but showing no memory of what just happened.

Yet by using population decoding we could show that small groups of neurons in the prefrontal cortex could remember everything: the choice just made, and milk delivered or not. Beautifully, we had shown that the legion of spikes in prefrontal cortex remembers even if each neuron seems to remember nothing; and even if the task to be solved does not explicitly require keeping something in mind. Because, after all, remembering decisions and their outcomes seems like a smart thing to do.

Which brings us neatly back to you: it's decision time. As we jump onto the spike leaving our neuron in prefrontal cortex, our spike and the spikes around us are holding memories of vital stuff. That there is a cookie, in the tatty cardboard box, on the brown desk abutting yours, lid open. That Adam is not behind you; Sarah is crossing the office, yet moving away, not looking in your direction. That Graham, hideous tie and all, is staring mid-distance, perhaps lost in contemplation of tomorrow's inexplicably scheduled lime and brown tie. That Idris and Kai, a few desks away, are immersed in conversation, bickering over the microwave cleaning rotation. Our spike and its compatriots have buffered all the information you need. Time to combine that information to make the crucial decision: to take the cookie, or not?

SOMETHING DOESN'T ADD UP

Many decisions require you to add up the evidence for each option. To accumulate the information flooding in about the state of the world, to weigh up which option to go for. If somewhere in your brain adds up evidence, then it must be where information about the world converges. The loneliest neurons in the prefrontal and parietal cortex seem perfectly placed to do that, too. And they are.[38]

We know this thanks to those random dots (figure 7.6). Remember the bored monkeys? Well they had to make a decision about which direction they thought the dots were moving in, and then move their eyes

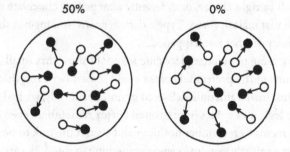

FIGURE 7.6. The randomly moving dots task. *Left*: In the task the
dots move across a circle, the movement of each dot indicated by
the arrow between its current (white) and next (black) position. The
player's job is to watch the dots and decide whether there is more
motion to the left or to the right. When lots of dots—here half of
them—move in the same direction, this is easy. *Right*: We can always
make no dots move in the same direction and ask for a decision
anyway, even though there is no correct answer. By doing this, we
can see what neural activity corresponds to the decision.

in that direction. They had just two options to decide between, that the
dots were moving left or moving right. And area MT, that region of
Highway Do filled with motion-loving neurons, is perfectly placed to
provide the necessary evidence:[39] if all the dots move in the same direction, then all its neurons liking that direction spike furiously, sending
lots of evidence for which direction the dots are moving in; if only a few
dots are moving consistently in the same direction, then only a few neurons liking that direction will get to send spikes, sending weak evidence
for which direction the dots are moving in, and accompanied by many
occasional spikes from other neurons responding to dots randomly
moving in their preferred direction. To accumulate this evidence into a
decision, somewhere downstream of area MT should be adding up
those spikes.

In the back end of parietal cortex and in bits of prefrontal cortex, both
targets of area MT neurons, that's exactly what we see.[40] Some neurons
in these regions increase the number of spikes they send while the dots
move in their preferred direction. And just as though they were adding
up evidence, the more the dots move in that direction, the faster the

increase in the number of spikes. Then, once the accumulating neurons for one direction have increased their activity enough, once they hit some threshold, the monkey looks in that direction. It decides: the dots are moving that way.

How do we know it's a decision? Because we can see errors, force a decision, and fiddle with causality. When the monkeys make errors, when they look in the wrong direction, activity in the accumulating neurons for that, wrong, direction hit the threshold first.[41] Sometimes we show dots all moving randomly, with no coherent direction, but ask for a decision anyway; whatever direction the monkey chooses, activity in the accumulating neurons from that direction hit the threshold first. And if we fiddle with causality by changing the number of spikes ourselves, that changes the decision. Stimulating a group of area MT neurons that prefer the same direction, forcing them to send lots of spikes even when they don't want to, consistently biases the decision in that direction. As though the accumulating neurons were simply adding up this extra evidence.[42]

This story seems like a Counter's dream, of single neurons sending messages to each other through the number of spikes sent, both evidence and its accumulation contained within the tick-tick-tick of a spike stream. But paradoxically it is in decision-making we see the full extent of the legion. If we turn off the regions of parietal cortex most rich in accumulating neurons, nothing happens.[43] Monkeys and rats go right ahead, making decisions just fine, with no measurable effect on how many errors they make, or how fast they make decisions, or how hard they find the task. Neurons in parietal cortex count evidence in their spikes just fine. But need have no effect at all.

For tuning does not mean function. Just because a neuron sends spikes in response to something, it does not mean that neuron plays a causal role in operating on that something. A neuron's spikes may respond to an edge, precede arm movement, elevate during a delay, increase to evidence, but not be necessary for any of those things. The brain is degenerate; it has multiple solutions to the same problem, multiple systems that can do the same job, but in different ways. And decision-making is where we find clear evidence of a degenerate system, with accumulating neurons distributed across large chunks of the brain, across bits of prefrontal and parietal cortex and crucial regions below the cortex,[44] yet most of which we can turn off individually with little to no effect on the decision itself.[45] Making a decision is so vital that the brain will try to find any way that it can to make one.

And when we look around us, the dominance of the legion in decision-making is no surprise, as these decision-making bits of prefrontal and parietal cortex are chock full of Type 2 dark neurons. Few neurons in the back end of parietal cortex exactly match the job description of the purely accumulating neuron, a neuron with a clear preferred choice direction, that is consistently activated by dots, and increments its activity to dots moving in its preferred direction.[46] Yet here the changes of spikes in many Type 2 neurons can be predicted in part by the motion of dots;[47] complex, jagged, weird changes—not simple, reliable increases or decreases in sending spikes as dots fly around a screen. So, perhaps unsurprisingly, by using our population decoders in bits of prefrontal cortex we can accurately decode the upcoming choice of the monkey, far better than from any one neuron.[48] And the decoding accumulates: the ability to predict the upcoming choice increases consistently from the start of dot movie until it's near perfect just before the monkey makes its decision. The legion decides.[49]

We grab a spike flowing out from the prefrontal cortex. Around us and in parietal cortex streams of spikes are adding up the evidence from your senses that there's a cookie and you're hungry and Adam is not at his desk and Angela and Ishmael are looking away and Ava's already had one, as have Zola and Dave and Shani and Hamid, and you're still hungry and there's nothing in your hand, and, and, and. All adding up to decide: take the cookie.

CHAPTER 8

A Moving Experience

READY, SET, GO

You decide to take the cookie. By the end of Highway What, the spike
we clung to was part of a legion identifying the objects before you as a
crumbly, oat-colored, ginger-strewn cookie in a "Cookies"-scrawled
cardboard box, identifying the people around you as Graham, Sarah,
Janice, Idris, Kai, a legion feeding all that information into your memory
buffer and your decision regions across prefrontal and parietal cortex,
so you knew what you were deciding about. By the end of Highway Do,
the spike our clones clung to was but one of a legion carrying messages
of the cookie's exact location (the abutting desk), size (big), orientation
(away from you), and movement (none, thank the stars; well, except for
the rotation of the Earth, but you can't see that in your reference frame,
because you aren't actually omnipotent despite what you claim). And
messages of where your office mates, delivery people, and assorted
hangers-on were standing, moving, and looking. Also all fed into memory and decision regions, so you knew where you were deciding about,
so you could gauge the feasibility of your surreptitious cookie-claiming
scenario.

Our spike is thrust onward to the motor regions of cortex, its arrival
part of the volley of messages meaning "move your hand and pick up
the cookie." Here the neurons are itching to complete the spike's journey down to the spinal cord, to pile spike upon spike into the motor
neurons, whose own spikes in turn tell your muscles to contract just so.

Seems simple. But from the spike's point of view, it is anything but. "Take the cookie" isn't a preprogrammed maneuver, a holistic action handled by dedicated cookie-grabbing neurons. That simple reach-and-grab is the coordinated contraction of muscles in your back, stomach, and side, to control your trunk as you lean across the desk; in your shoulder, to rotate it; in your upper arm, the triceps, to extend it; in your forearm, to open your hand and your fingers, to open them, extend them just the right amount to land on the cookie's crumbly edges. And then reverse: contract your fingers to grasp the cookie, as your arm muscles contract to pull your arm back toward you, forearm and shoulder rotating to swing your hand up to your gaping maw, as a complex sequence of relaxing and contracting in your back, stomach, and side returns you from leaning into the snug comfort of the black and neon green mesh office chair.

How do neurons in the motor bits of cortex do all that? For reaching and grasping, those crucial steps toward the prized cookie, the paths are well mapped out.[1] To complete our journey, we will need to follow our spike as it traverses the reach-planning regions of parietal cortex, through the premotor cortex where the reach is prepared, and on to the primary motor cortex, the ultimate source of most neurons that send axons directly to spine. Launching on a spike from a neuron in the parietal reach region, we're part of a legion carrying messages of where exactly the cookie is located and that it is, in fact, a flattened dome of tastiness.[2] We land on a layer three neuron in the premotor cortex, one who is quite literally preparing to make you move. For here a surge of spikes are sent in the few hundred milliseconds before your arm starts to move.

As I foretold in the last chapter, Counters place movement neurons among their favorites. Indeed, if we hung out here in premotor cortex and counted spikes from neurons around us during this preparation to move, some would show tuning to the imminent arm movement, to the imminent patterns of contraction and release of single muscles; others would seem to show tuning for more abstract parameters, to the imminent direction or velocity or the intended end position of your arm (respectively: forward, quickly, the cookie).[3] But many, perhaps most, neurons around us have no such tuning. More confusing is that many of those with some tuning during the preparation to move seem to change that tuning during the actual movement.[4] Which makes no sense if the job of those tuned neurons is to make happen the thing they were tuned for. And on our voyage what have we learned about such a messy mélange of individually incomprehensible single neurons? Exactly—the message is in the legion of spikes from many neurons. But what though is that message when preparing to move?

Movement poses a new problem, one we've not yet encountered. Moving your arm takes time. So the spikes to move your arm and hand have to unfold in just the right way over time, contracting the right muscles in the right sequence. And, once initiated, once the movement has started, they need to continue unfolding, need to be self-sustaining. You know how you reach for a pen, only for your arm to suddenly flop down to your side halfway through? No, you don't, because it's never happened. Once the spikes start to make your arm move, they carry on until that move is done, come hell or high water. Making such self-sustaining spikes needs many neurons, wired together—we'll revisit this crucial idea later in this chapter and the next. But before the legion of neurons can emit the self-sustaining spikes in the correct sequence over time, they first have to get themselves to the start of that sequence. And that's what these mysterious spikes during the preparation seem to be: the spikes in premotor cortex moving its own neurons and the neurons in motor cortex into the right starting point to then execute the right sequence of spikes—and the correct set of muscle commands.[5]

Hang on a second. If we stimulate "arm" neurons in motor or premotor cortex, the arm moves. Yet when we prepare to move, there are lots of "arm" neurons in those motor bits of cortex sending lots of spikes—and the arm doesn't budge an inch. If these spikes always meant "move arm," we'd be eternally flapping wildly about like a toddler fending a wasp off a precious ice cream. One of neuroscience's big mysteries is how muscles know when not to do anything. Why are we not always flapping our arms about?

The answer lies in the newly discovered "null space." This is exactly as sci-fi as it sounds. If right now we followed a spike out of premotor cortex down to the spinal cord, we'd enter an alternate dimension where it has no effect on the world. For this dimension is a cunning arrangement of spikes across the neurons in the motor bits of cortex in which the spikes signaling an arm (or leg, or hand, or neck) movement are kept in balance; increases in some neurons' spikes are counteracted by decreases in others, so the total number of spikes remains about the same. And because the sum remains the same, the motor neurons in the spine do not change their output. And because the motor neurons do not change their output, in turn the muscles they target do not change how much they contract. Lots of spikes, no movement. The null space is the space of all possible ways the neurons in the motor bits of cortex controlling a body part can add up to the same number of total spikes.[6] Yet, all the while each of those neurons is being moved into the right starting point for the movement sequence.[7]

Great, you're thinking, we're prepped. Now let's jump a spike heading to motor cortex, move the arm, grab that cookie, and we're done. Ah, but if only it were so simple. For now we must pitch ourselves into the undignified scrap for control of your body.

WHAT DO I DO NOW?

For your brain there's a far grander challenge than just moving your arm: how do you know it is safe to move your arm now? Something else more important could already be happening, or need to happen instead. Be it you're in flip-flops sprinting ungainly away from an angry squirrel, or you imminently need to hit the high F in "Let It Go" thanks to a drunken sign-up to the X Factor auditions, the last thing you need to add to your deep regret is to spasm your hands randomly in the air.

And that's why our spike is at the same time sent down to the basal ganglia, the sullen outcropping of neurons underneath the frontal cortex, to ask: can I move my hand now? Our work and the work of many others has shown how the basal ganglia are the brain's stern parent.[8] Endless spikes pouring out of the basal ganglia are constantly stopping you from doing what you want. They inhibit everything they touch. No, you can't do that. No. No. No. In order to move your arm, our spike must navigate the twisted pathways of the basal ganglia to reach their output and momentarily turn off the endless stream of spikes.

We land first in the striatum, gateway to the basal ganglia. The route we took here is one of uncountably many possible ones: first a jump between the pyramidal neurons of layers three and five of premotor cortex, then, instead of taking the branch of axon through the white matter and on to motor cortex, this time we cling to the cloned spike that turns down the branch to the striatum. Among the other routes, next door to us in layer five was the type of pyramidal neuron that sends its axon all the way to the spine; it too dispatched a cloned spike to the striatum. Indeed, all layer five neurons that send axons within the cortex seem to send a branch down to the striatum. And all layer five neurons that send axons destined for the brain stem or the spine also dispatch a clone of every spike to the striatum. Which means that as we arrive on a spike from premotor cortex, we're joined by millions of other spikes, coming from all over the cortex, from all over the prefrontal and parietal cortices, the memory buffers, the evidence accumulators, and more, from all parts of motor cortex, from all types of sensory cortex, from touch, sound, and many stops along Highway What and Highway Do. All sending information about what is in the world, and what could be done about it. Striatum knows all.[9]

And we have compelling evidence that the striatum uses that knowledge as votes for different courses of action. Electrically stimulate a small group of neurons in the striatum, and you'll make a body part move.[10] More precisely, stimulate a group of cortical neurons sending axons to the striatum, and you'll bias behavior toward whatever those neurons encode; for example, stimulate neurons of the auditory cortex encoding high-frequency sounds, and the mouse will check for reward at the location previously predicted by high frequencies.[11] Turn off the striatum, and you'll deeply and permanently impair the mouse's ability to make the correct choice of action.[12]

Some of our most compelling evidence that the striatum controls the choice of action comes from we humans. Striatum malfunction is at the core of most of our movement disorders. The most striking outward signs of Parkinson's disease are its problems with moving—the rigidity of the body, the slowness of movement, the inability to start moving. The death of dopamine neurons prefigures these symptoms, and by their death the striatum loses its source of dopamine. Remove dopamine from an animal's striatum, and Parkinson's-like movements result. In Huntington's disease, the death of the striatum's principle cells prefigures that disease's uncontrolled, thrashing limb movements. And more: dystonia, with its unnatural long-lasting muscle contractions; Tourette's syndrome with its tics and problems of speech control; even disorders of inappropriate actions, such as obsessive compulsive disorder. Malfunctions of the striatum are implicated in all of them; all are disorders of making the right choice of action.

Not just making the right choice, but stopping actions too. Flowing out from the striatum are two pathways of axons, from two groups of its principal neurons (figure 8.1). One pathway sends axons directly to the output neurons of the basal ganglia. This direct pathway selects action. Stimulate just its neurons, and depending on what it was already doing, an animal will start scampering, or start a sequence of moves, or insert a new action into an ongoing sequence of moves. The other pathway follows a more torturous, indirect route to the output neurons of the basal ganglia, via jumps onto neurons in the internal nuclei of the basal ganglia. This indirect pathway controls and aborts action. Stimulate just its neurons, and depending on what the animal was already doing, it will stop scampering, or fail to start a sequence of moves, or abort an ongoing sequence of movements. Together, these direct and indirect paths out of the striatum exercise tight, competing control over what you'll do next.[13]

Yet for all its barrage of incoming information, its crucial role in so many disorders, and its competing neurons, the striatum says very little. The striatum is massive, with about one-fifth the number of neurons in your entire cortex, but is deathly silent. Hook up the lab speakers to an electrode as you lower it through the layers of cortex, and you'll hear the constant chatter of spikes, the tick-tick-tick as the electrode descends; suddenly, as the electrode bursts through white matter and into the striatum, the lab is becalmed, the speakers fall silent. The principal striatum neuron can absorb a fantastic number of spikes without

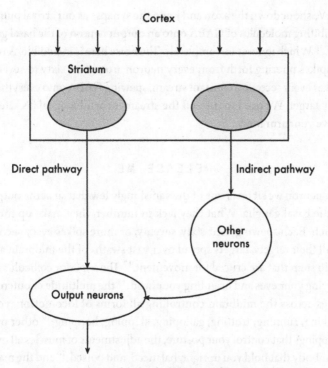

FIGURE 8.1. Basic layout of the basal ganglia. Axons from all over the cortex send spikes streaming into the striatum. The striatum divides these among two groups of its neurons: the direct and indirect pathways to the output neurons of the basal ganglia.

making a new one. I once estimated the principal neuron needs upwards of 500 excitatory spikes in one second to make one new spike, fivefold more than needed by a pyramidal neuron in the cortex.[14] Indeed, the principal neurons seem designed precisely to be choosy, to ignore anything but a concerted volley of spikes from the cortex,[15] perhaps to filter out the noise, to make sure that random smatterings of spikes from the cortex don't invoke an unwanted, inappropriate, or downright dangerous action. Fortunately for us, we're on the leading edge of just such a volley converging from premotor cortex. So we wait but briefly on the principal neuron's body until we grab a spike on the direct route out.

We shoot down the axon and jump the synapse as our arrival pumps inhibiting molecules of GABA onto an output neuron of the basal ganglia.[16] Which ignores us completely. The noise here is incredible. A roar of spikes pouring forth from every neuron around us, sixty to seventy spikes every second, a constant stream, spewing GABA onto everything they target. We need to turn off the stream, to send a signal it's safe to move your arm now.

RELEASE ME

The neuron we sit on is one of the vanishingly few that sit at the output of the basal ganglia. What they lack in number, they make up for in punch. Each spewing their sixty, seventy, or more spikes every second to all their targets, targets spread over vast swaths of the midbrain and brain stem that are crucial for movement.[17] The superior colliculus for moving your eyes and orienting your head;[18] the multitude of outcroppings across the midbrain controlling all forms of locomotion, your walking, running, trotting, galloping, skipping, hopping;[19] other outcroppings that control your posture, the adjustments of muscles all over your body that hold you upright, balanced, and poised;[20] and the many subdivisions of the thalamus, the gateway back to the cortex.[21]

The neurons in all these regions are permanently drowning in the GABA spewing from the basal ganglia output neurons, GABA that suppresses any upward blips of voltage, that suppresses the voltage of these target neurons from reaching their tipping point and sending a spike. GABA that suppresses movement.

Turning off this torrent of GABA, releasing the target neurons from their suppression, allows movement. Switch off the output neurons in rodents, and their target neurons in thalamus immediately start sending spikes.[22] If you were to permanently switch off the output neurons that control where your eyes are looking, then you could not stop your eyes from constantly looking at new things.[23] The reverse appears to happen in Parkinson's disease: turning off the torrent of spikes from the output neurons becomes ever more difficult, and movement slows or freezes completely.[24] So we need to turn off the torrent so you can move. (And we want to turn off just those output neurons that control the specific movements you want to make now—the adjustment of posture as you lean forward, the eye and head orienting to the cookie, the arm reaching to it.)

We already know the striatum's direct pathway is the key: it selects actions, so must be able to turn off the torrent of spikes from the output neurons. Indeed, the striatum is an inverter of the cortex, turning excitation from cortical neurons into inhibition. Our arrival on a spike from striatum sent GABA to the output neuron on which we sit. But striatum is silent, most of the time. How can neurons sending so few spikes turn off this huge wave of spikes from the output neurons?

The brain uses the trick of scale to solve this problem, a shining example of how dark neurons can do useful work. The striatum's neurons outnumber the output neurons by two orders of magnitude; in the rat, by three million to about thirty thousand.[25] Even if each striatum neuron only contacts one hundred output neurons, and likely they contact many more, then each output neuron will get ten thousand inputs from the striatum. All ten thousand sending inhibition to that one output neuron of the basal ganglia. So it just needs a small fraction, perhaps just 1 percent of those inputs, to send one or two spikes, and hundreds of GABA receptors activate on a single output neuron, shutting down its spike torrent.

We arrived on the leading edge of a massive volley of spikes from the striatum, a volley that now piles into this output neuron and those around us. The volley builds, the GABA flows freely, the inhibition accumulates—and the torrent of spikes from this neighborhood of output neurons begins to dip, slowly at first, then faster, until some stop completely.[26]

Success! But success is bittersweet for us. It means we're about to hit a dead end—we're about to quash the spikes coming out of the neuron we're sitting on and be left marooned here while your arm reaches out for the confounded cookie. We leap onto the last spike out, follow it up the axon toward the motor regions of the thalamus, while its clones follow branches down into the midbrain and brain stem, the last hurrah of inhibition, releasing the neurons needed to lean, twist, move the shoulder, extend the arm. Into thalamus, and jump the gap. As the GABA-driven downward blip of voltage our spike induced fades, this thalamus neuron is springing into life, its voltage rocketing upward now the torrent of GABA has dissipated. A spike is born; we hop on, and ascend back to the motor cortex.

WITHIN YOUR GRASP

Just in time too. With the extra oomph of spikes from the thalamus now arriving across the motor regions of cortex, the preparation is complete. As we arrive, following the molecules across the gap onto a pyramidal neuron in layer three of motor cortex, the neurons around us are ready to start unfolding the sequence of spikes to reach your hand and grasp the prize. And while we could check if each of the neurons around us was tuned for a specific muscle contraction, or velocity, or parameter of movement—and many people have over the years—we know by now that most neurons will not have any tuning.

Motor cortex was perhaps the first to reveal to us the power of the legion of spikes. In 1986, Apostolos Georgopoulos and colleagues showed that we could combine the spikes from a small population of neurons in motor cortex to accurately decode the direction an arm was moving in three-dimensional space.[27] But to do this, they only considered neurons tuned for direction, combining the spikes of neurons that each had a clear preference for a particular direction of movement.[28] It took many more years to realize we could just as easily decode movement from any old collection of neurons in motor cortex, that tuning was not important, that the legion was key.[29] Now we can even use the spikes from about a hundred neurons in motor cortex to decode which of twenty different grips is being used to hold an object.[30]

But, remember: movement is a continuous thing that unfolds over time, an evolution of muscle contractions at different moments, driven by self-sustaining spikes in the right sequence. And now the preparatory activity has moved the motor cortex neurons to the right state to start the sequence that will move your arm. That sequence is disarmingly, charmingly, weirdly simple.

Mark Churchland, Krishna Shenoy, and their collaborators have reported that, during arm movements, these self-sustaining dynamics in motor cortex have a simple, consistent behavior: they rotate.[31] While each neuron can have a complex-looking sequence of increases and decreases in spikes while an arm is moving, collectively those complex sequences trace out an arc, a reliable ebb and flow of spikes across all the neurons, a relay race of whose turn it is to send their spikes now.

This simple shape of the legion's spikes contrasts with what the muscles are doing. Under the apparently smooth movement of a reaching and turning arm are occasional sudden changes in the contractions of the muscles, yet we do not see sudden changes in the spikes from most neurons in the motor cortex, no sudden veering off course, wild jumps of sudden bursts of spikes, or sullen silence. Rather, while your arm is moving, the motor cortex is dominated by smooth changes in the number of spikes being sent across its neurons. Indeed, when your arm is itself rotating, like turning a handle, its muscles are beset by sudden changes, rapidly contracting and releasing, yet in motor cortex the spikes are tracing what is basically a circle for each rotation of the handle.[32] This is exactly what you'd expect to see if most of the spikes are not sending commands to move at all, but are instead to create self-sustaining spikes. The lack of tuning in most neurons of the motor cortex is then because most neurons are producing the self-sustaining spikes, to keep movement going.

The command to move the arm is still there in the spikes being sent around us. When the arm is rotating a handle, we can directly decode the muscle contractions from ripples in the circle traced by the legion of spikes.[33] We can predict the trajectory of a reaching arm from the shape of the arc traced by the legion of spikes.[34] We can even decode arm movements from the changes to the legion's spikes that are common across different tasks.[35] Which all means that around us wafts a wave of spikes, most driving a smooth familiar pattern, keeping the movement going, some unique, those needed to contract a particular muscle just so, right now.

Time to find out what this waft of spikes is talking to. Grab that spike, quick. We descend from layer three to layer five again, this time careful to hop onto a pyramidal neuron with a big, thick axon, one that will take us across the epic distance from layer five of your motor cortex to the upper part of your spinal cord, where live the motor neurons for moving your arm. Into the pyramidal tract we launch, a dense parallel bundle of axons, all streaming from layer five neurons down toward the spine, all now lit up with spikes.

Many of these axons are on their way direct to the spinal cord. Those from the arm and hand regions of the motor cortex will stream down toward the outer edge of the spinal cord, to hit the motor neurons whose axons snake out toward the muscles across the arm. Those from the shoulder and trunk regions of motor cortex stream toward the spine's center. On the way down, many axons will send branches into different parts of the brain stem, launching clones of spikes into small pockets of neurons crucial for coordinating your lean across the desk, the shifts in posture and balance; crucial for coordinating precisely how much each muscle has to contract to hit the target and pick it up;[36] and crucial for you to hold your breath momentarily, placing your body on silent running, minimizing the risk of drawing attention during the cookie grab.

We've reached the spinal cord in the blink of an eye, a few milliseconds, that fat, myelin-wrapped axon of the layer five neuron an epitome of why the brain uses spikes to send messages far and fast. As we plummet through the top few segments of the spinal cord, we're surrounded by another complex network of interconnected neurons, some types that excite their targets, some that inhibit them, both of which send their spikes to the motor neurons, the ultimate conveyors of spikes to the muscles.[37] Such a circuit of inhibitory and excitatory neurons converging on each other is strongly reminiscent of the cortex. Researchers in Rune Berg's lab in Copenhagen have shown this reminiscence is deep,[38] so deep that the motor neurons fire random-looking, irregularly timed spikes; that they do so because the excitatory and inhibitory inputs arriving at them are kept in exquisite balance; and that the number of spikes sent by each neuron follows that "long-tailed" distribution, most neurons firing less than one spike every second, a few firing tens of spikes. Even here, at the very last step of our journey, outside the brain proper, we find the legion at work around us.

Most axons from the motor cortex land within the spine's network of neurons, targeting the types of neuron that ultimately connect to the right motor neurons for the movement that is to be made.[39] But we're flying first class, following the axon from cortex that controls the fingers. We land directly on a motor neuron. This route from brain to spine is only available to us higher primates, seemingly the crucial ingredient in our extraordinary ability to manipulate the world with our hands, for the fine precision control of our fingers.[40]

Gliding across the gap onto the motor neuron's dendritic tree, for one last time we flow with the little voltage blip down toward its body, our blip and others driving the motor neuron to its tipping point, hold tight to the spike as it zips through its long snaking axon, up into the shoulder, down the arm, and into the endplate, a synapse onto the flexor digitorum muscle, watch the molecules flow and lock into receptor, feel the muscle contract—and your finger touches the cookie. We're done.

CHAPTER 9

Spontaneity

THE WORLD IS NOT ENOUGH

Everywhere we've been, spikes were already there. In the motor cortex we entered the null space, traveling on a spike that simply added to spikes already there—contracting the muscle needed a change in the number of spikes constantly streaming from the motor cortex. In prefrontal and parietal cortex, when we landed our spike on a neuron, spikes were already pouring out of it and the neurons around us, buffering memories and accumulating evidence. You may recall that at every jump along Highway What and Highway Do, spikes were streaming back past us, back toward V1, spikes that already existed throughout the visual regions of cortex before you'd seen the cookie.

And in V1 itself, spikes were already there. Departing the retina, our first landfall in your cortex was the simple cell in layer four. Yet descending its tree to its body, we discovered we were but one of many voltage blips arriving there, blips caused by spikes from other neurons in layer four of your visual cortex, including those pesky interneurons spewing GABA. Indeed, the spikes arriving from the eye were heavily outnumbered: on a simple cell about 5 percent of its inputs come directly from the eyes.[1]

Everywhere we went, spikes were already there. Yet we were in the first wave of spikes sent by the cookie falling on your retina. How can that be? What made those spikes?

To explain that means dismantling two deep misconceptions about the spike. I held them, and most neuroscientists do too. But now at the end of our spike's journey we have seen enough to know better.

The first misconception is that all spikes are caused by events in the world. That if we see a neuron sending a spike, then it must be linked to something happening in the world. That, for example, a spike in motor cortex ultimately has its origins in a spike from the retina caused by seeing something.

This is not true. Many spikes, possibly most spikes, have no cause in the outside world. Collectively, we call these spontaneous activity—spikes that arise seemingly unbidden.

One explanation for these "spontaneous" spikes is that they are the traces, the echo, of what just happened a moment before. That perhaps on our journey these "spontaneous" spikes were created by the initial glimpse of that scruffy cardboard box with its half-open lid, before your eyes fell on the cookie's form. A tempting explanation, but incomplete. We know they do arise unbidden. There have been spontaneous spikes in your brain whatever you've been doing, throughout your entire life.

Close your eyes. No light enters your eyes, nothing needs sending to the vision parts of your cortex. You'd think the spike would take a well-deserved rest at this point. But no. Your visual cortex is ceaselessly spiking, eyes closed or eyes open, whether there is anything to see or not. Indeed, brain imaging has shown us that a whole network of cortical regions are paradoxically at their most active when you're quietly resting, eyes closed; when you do something, this network reduces its activity. The spontaneous activity in this "default" network is not only no echo of the outside world, but is also reduced by engaging with it.[2]

Now sleep. Naively one might think that sleep is your brain "turning off," that neurons stop talking to each other. Brains can't do this; after all, neurons not talking to each other is a key element of many countries' legal definition of death. Instead, neurons all over your brain are sending spikes throughout your slumbering.[3] In the deepest stage of deep sleep, neurons all over your cortex switch about once a second between a burst of spikes and a burst of silence. And they switch together, synchronized across your cortex, a coordination of spiking then silence a world away from the irregular, desynchronized firing of the alert brain. Yet cortical neurons send just as many or more spikes in total in this slow-wave sleep as when you're awake. In REM sleep, the same cortical neurons send irregular, uncoordinated spikes, looking for all the world like you are awake. Except they have little effect on the world, because the motor neurons in the spine are inactive, profoundly inhibited, blocking the brain's access to your body muscles.[4] Sleep, then, is filled with spontaneous spikes, born of no inputs, yet filling the brain with the buzz of activity.

That buzz of spontaneous activity has been there since your first days on Earth.[5] Everywhere in the developing brain, in the retina, in the cortex, in the striatum, in the dark recesses of the midbrain and brain stem, neurons send spikes spontaneously before there is anything to cause them. Spikes in the retina and the visual cortex before the eyes open. Spikes in the whisker bits of sensory cortex before the whiskers move.

The spontaneous activity of the developing brain has many possible roles. It controls how many neurons are born, and which die. It shapes the neurons themselves and the wiring between them. The initial connections between neurons in the cortex seem to be made at random, synapses appearing wherever axon touches dendrite.[6] How far a neuron extends its dendrite as it develops is not fixed, but is regulated by its own activity. The tree grows if the neuron is not active enough, seeking new inputs, a bigger legion to drive the neuron to its tipping point more easily. Conversely, the tree shrinks if the neuron is too active, trying to lose inputs, shrink the legion, and make it harder to reach the tipping point.[7] Spontaneous spikes have been there since the beginning.

Spontaneous spikes no doubt have different roles to play in development, in sleep, and in your moment-to-moment waking life. But regardless of their role these spontaneous spikes have in common the same two options for their creation: from the neuron itself, or from the circuit.

WHO SAID THAT?

Here we meet the second of our misconceptions about spikes: that neurons always need inputs from other neurons to make a spike. That if we see a neuron sending a spike, then it must have been caused by a cascade of voltage blips down its dendritic tree, combining at the neuron's body to drive it to the tipping point for a spike.

This is not true either. For the most straightforward way to get spontaneous spikes is for a neuron to make its own spikes.[8] To be literally spontaneous.

We ran into some of these crazy neurons down in the basal ganglia. That wall of unending GABA from the output neurons was spontaneous. Their ceaseless output, stopping you from dancing the polka on your boss's desk, comes from neurons that are so committed to making spikes they'll even do it sitting alone in a dish. Cut an output neuron of the basal ganglia from the brain, and for as long as you can keep it alive it will spit out a spike with clockwork regularity every 100 milliseconds or so.

As we learned at the start of our journey, a spike is born from a rapid opening-and-closing sequence of holes in the neuron's skin. Holes of two types: those that let sodium rush in; and those through which potassium is pumped out. This quick-fire sequence is triggered by the neuron's voltage reaching its tipping point. So for a neuron to make its very own, spontaneous spikes, without input from other neurons, its voltage must reach the tipping point all by itself. To do this, neurons that make their own spikes are equipped with special types of holes in the neuron's skin, holes that create a feedback loop; as the neuron's voltage plummets at the end of a spike, these special holes slowly open, letting positive ions slowly accumulate in the neuron, driving up its voltage back toward the tipping point. Which births a spike, and starts the whole process anew. This is exactly the same mechanism that drives the constant pulsing of the pacemaker cells in your heart, whose spontaneous, tick-tock spikes keep your heart beating and you alive.

We find pacemaker neurons throughout the brain. They're in all parts of the basal ganglia, not just its output: the cunningly named subthalamic nucleus (translation: bit of brain tucked under the thalamus) and globus pallidus (translation: the pale globe bit) are entirely made of pacemaker neurons; the striatum is peppered with a giant pacemaking interneuron.[9] Vitally important are the swath of pacemaker neurons across your midbrain that transmit neuromodulators, neurons whose unending spikes deliver a fresh supply of serotonin, noradrenaline, and dopamine all over your brain. You already met an example of how they are vitally important: losing the constant supply of dopamine to the striatum we know as Parkinson's disease. Pacemaking neurons also appear widely during the very early development of the brain, across the cortex, in the retina and elsewhere, driving the spontaneous activity that guides the growth of neurons and the creation of connections.[10] While many of the pacemaking neurons in the cortex disappear shortly after birth, there are even reports of rare pacemaker neurons in the developed cortex of mice.[11]

Beyond these pacemaking neurons able to entirely self-generate spikes are other types of neurons, verbose, chatty neurons, that generate more spikes than asked for.[12] Some verbose neurons create a spike after being released from inhibition. They bounce back, rebound, from suppression and announce to the world "I'm free!" Some verbose neurons send a whole burst of spikes after reaching the tipping point for the first spike, driven by special holes in the skin that are popped open by that first spike and allow positive ions to accumulate for a while, pushing the neuron's voltage repeatedly back to its tipping point for a spike. The cortex has some of these verbose neurons too, neurons that have many ways of speaking without prompting.

But spontaneous spikes born of no input create unstructured activity, with each neuron doing their own, independent thing. And, with the possible exception of the very early developing brain, the pacemakers are not enough in number to create spikes across a whole network of neurons. The rare pacemakers and bursters in the cortex are not enough to generate the spontaneous activity we've seen throughout our journey. No: the source of most spontaneous activity is the network itself.

FEEDBACK

Take a slice of the cortex, and drop it into a dish. Now record lots of its neurons. Despite being connected to nothing but each other and receiving no input from the outside world, the slice will be abuzz with spikes, dotted with groups of neurons that fire spikes together.[13] Bathe that slice in a chemical soup that mimics the salty water sitting outside neurons' skins in the intact brain, and neurons across the whole thing will spontaneously fall into the slow-wave rhythm, each alternating a burst of spikes and silence every few seconds.[14] A deep-sleeping slice of isolated cortex. Slices of the hippocampus not only spontaneously spike in coordinated patterns but also spontaneously change those patterns every few minutes.[15] Yet bathe these slices in a chemical soup that blocks the synapses between neurons and almost all the spikes vanish, leaving just the rare pacemakers sputtering away. In all these isolated bits of brain, the vast majority of spikes are generated by the network of connections between the neurons.

The key is feedback. Circuits of neurons wired together can sustain their own activity, by feeding back spikes to each other, creating new spikes in neurons solely from spikes already traveling around the circuit. We know what the brain needs to create such feedback: it needs a legion of spikes to make one new spike, so the feedback must be from many neurons; and excitatory feedback by itself runs away, so we need inhibition to balance the feedback excitation. Our journey showed us these circuits exist all over the cortex.

We saw that a pyramidal neuron sends much of its axon nearby, branching furiously to connect to thousands of its neighboring pyramidal neurons. As intrepid, impatient explorers shooting the rapids, hopping spikes from one neuron to the next, we explored but one of those connections. If we'd been patient cartographers, dutifully charting the possible routes from our starting point, we would have discovered each pyramidal neuron sits at the center of myriad feedback loops.

Start on one pyramidal neuron and follow its axon to a neighbor. Then follow that neuron's axon to one of its neighbors. And keep on going, tracing a chain of pyramidal neurons. There is always a chain that ends up back at our starting neuron. It could be an axon sent directly back from the very next neuron, giving immediate feedback; it could be three, five, or ten neurons later. But we can always find a complete loop back to the start. So by sending a spike, a pyramidal neuron creates the potential to excite itself in the very near future.

The potential, because most loops will fail to feedback that spike to the starting neuron. At many neurons in many of those loops, there won't be enough other upward voltage blips arriving at roughly the same time to make a spike. And even if a spike is made, we know it is likely to fail at a crucial synapse, so breaking the chain. Yet we can guarantee the spike will find a way back, because the numbers of loops are astronomical, almost uncountable.

Imagine a pyramidal neuron in a tiny neighborhood of just ten thousand other neurons, connecting to each neuron in that neighborhood with a probability of 10 percent. Then it will have about one hundred immediate feedback loops, direct feedback from the neurons it directly contacts: our starting neuron will connect to about one thousand of the ten thousand; of those one thousand, 10 percent will connect back to our starting neuron, giving one hundred immediate feedback loops. But each of those one thousand contacted neurons will each in turn contact about one thousand further neurons, a million neurons now in total, who could each connect back to our starting neuron. Run the math, and our starting neuron will have about ten thousand feedback loops that are two neurons long, and about ten million feedback loops that are three neurons long.[16] So in your cortex, with vastly more than ten thousand neurons in each neighborhood, there will be more than enough loops conveying spikes from start to finish, feeding back to create a rapid series of voltage blips in the original neuron, driving it to spike. Starting the whole process again.

(And that's just within the local neighborhood of neurons within a single layer of the cortex. Zooming out, we see more feedback, more ways to sustain spikes. There are loops between layers within the same region of cortex.[17] In your visual cortex, we sprinted from layer four up to layers two and three, then down to layer five. Painstaking cartography would have shown us that we could loop back to layer four from layers two and three, or loop from layer five back up to layers two and three. There are loops between regions of the cortex. Just like we saw on both visual Highways, with spikes streaming back past us toward the regions—V1, V2, V4, MT—we'd just left.[18] And there are long loops that leave the cortex, run through the thalamus and back to cortex again.[19] Loops near and far.)

That's the legion for making spontaneous activity; now what about the necessary balance? We've seen that too. On our journey we also saw that a pyramidal neuron's axon contacts some of those rarer interneurons that transmit GABA and inhibit their targets. Rare indeed: about 90 percent of the inputs to a cortical neuron excite it; only about 10 percent inhibit it. But we saw they were powerful. Positioned close to the neuron's body, these GABA synapses annihilate excitatory blips trying to get past. And those GABA synapses have a much lower rate of failure than excitatory ones. So spikes from these interneurons are reliable and powerful. Some interneurons will connect directly back to the originating pyramidal neuron. Others inhibit pyramidal neurons farther along the loop, stopping those from spiking. Yet others will be at the end of the loop, ultimately sending inhibition back to the starting neuron. Thus one spike from a pyramidal neuron fires a starting gun on millions of races through the network, excitation and inhibition flowing through the dense web of intertwined downstream neurons, ultimately creating a flow of excitation and inhibition back onto that starting neuron.

From deep theoretical work, we know the fine details of such a network of neurons will determine both the kind of spontaneous activity and how long it lasts. The fine details of the exact mixture of excitatory and inhibitory neurons, of what is wired to what, and how strongly.[20] Some wirings can massively extend in time the network's response beyond an initial short, sharp input.[21] Or completely self-sustain after that input. Or even have a network self-sustain spikes without any kick-start input at all, because the spikes already within it are sufficient to make more spikes for as long as that network is intact. Some making irregular patterns of spikes, some making oscillations, spikes coming in waves, and some collapsing into chaos.

And theorists have a good handle on what kind of networks drive the self-sustaining spikes we saw in action in your me-want-cookie brain.

Most obviously, the self-sustaining activity in your motor cortex that drove your arm up and across the desk, fingers stretching to grasp the cookie. There's a well-developed theory of the type of self-sustaining networks that make the "arc" of firing across the legion in motor cortex.[22] A theory that says the network of the motor cortex has the feedback loops, balance, and wiring such that when prodded with an input, it unfurls a self-sustaining sequence of neurons firing spikes. Self-sustaining, but transient—the sequence of firing long outlasts the initial prod, but disappears after tens of milliseconds.

Indeed, from humans to mice to sea slugs to the maggots of vinegar flies, movements of all kinds are created by such self-sustaining networks of neurons.[23] Everywhere we find rhythmic movements—walking, crawling, swimming—we find circuits of neurons that self-sustain repeated activity. That once prodded into life, self-sustain neurons firing cycles of spikes then silence. Each burst of spikes from multiple neurons at the same time driving the contraction of a muscle; each cycle of spikes across all the neurons creating a single repetition of the movement, one stride, one push-pull of the crawl, one stroke of the swim.

Your prefrontal cortex neurons that sustained memories of the cookie, the box, and the personnel in your office did so in their persistent spikes. And persistent spikes need sustaining. So our best theory for this memory buffer is that prefrontal cortex contains networks of neurons that feedback on each other, that given a prod of input—a tatty box lid, a fleeting glimpse of a coworker—self-sustain spikes holding that input in mind.[24]

The same kinds of network are likely what support decision-making too. The accumulated spikes in parietal and prefrontal cortex can last for seconds. And lasting for seconds implies the spikes must be sustained by the neurons themselves. So our best theories for how neurons send spikes throughout long-gestating decisions is that they too are part of a network feeding back on each other.[25] One set of theories places this feedback between neurons in the local circuits in cortex;[26] another places the feedback across the loop to and from the cortex, running through the basal ganglia and thalamus.[27]

So we think feedback networks are everywhere in the cortex, networks of neurons that create their own dynamics. In motor, prefrontal, and parietal cortices we see the kind of long-lasting spikes that are the signatures of self-sustaining networks. Elsewhere in cortex, even when we haven't explicitly looked for these kinds of self-sustaining spikes from neurons, we can see the blueprint of these feedback networks. The wiring between neurons in the cortex is roughly the same everywhere, even in the first regions of visual cortex. There are myriad feedback loops to pyramidal neurons in local neighborhoods. So even if we cannot unambiguously see self-sustaining spikes in these bits of the cortex, it seems likely these regions are capable of producing them.

To me, this raises an interesting idea. In your motor cortex we saw many neurons with no apparent tuning to the movement of your arm. To the researchers who discovered this lack of tuning this came as no surprise, because the theory that motor cortex contains a self-sustaining network predicts it: there should be neurons with no tuning, because those neurons were part of feedback loops, not directly getting input or sending output directly to the spinal cord. But on our journey through your brain we gave a name to neurons that had no obvious response to the outside world: Type 2 dark neurons. We saw Type 2 dark neurons everywhere.

The idea then is simple: these Type 2 dark neurons are actually telling us that all cortical regions are dominated by self-sustained dynamics. That in fact Type 2 dark neurons, those "active but not tuned" neurons, are all there to generate the spontaneous activity.[28] In V1, V2, V4, throughout Highway What and Highway Do, everywhere we have (and haven't) been. Everywhere in the cortex we think we've not seen spikes from sustained dynamics, perhaps we've been looking at them the whole time—they just are the spikes of Type 2 dark neurons.

We see spontaneous spikes all over the cortex. And as just recounted, we have a good handle on how they are generated. But in the awake behaving brain what are they for? Both pacemaking and network forms of spontaneous activity are a neuroscientist's nightmare; they use an extraordinary amount of energy, yet being born of no input, they seem to carry no message about the world, have no code. Solving the deep mystery of their existence is the topic of the final chapter.

CHAPTER 10

But a Moment in Time

The cookie is on its way toward your mouth. As it ascends, spikes have been all over your brain: from the retina, to visual cortex, through the prefrontal cortex, onward to motor cortex and the basal ganglia, and down through the brain stem and on to the spine. For you, it was but a moment in time. A mere two seconds. From looking to reaching, the blink of an eye.

From the simplest reflex, snatching your hand back from a lava-hot coffee, to freezing when a tiger coughs behind you, to deciding whether the black and white blur streaking toward you is an ecstatic Dalmatian or angry panda, to your brain filling in the next words in your favorite song so you can belt out the chorus flatter than Denmark, the speed with which your brain can—must—respond to the outside world puts brutally tough limits on how spikes work. And your brain can react obscenely fast.

Simply responding to the outside world takes it just a handful of milliseconds. I flash you a picture and the first detectable change in the spikes from your retinal ganglion cells happens about 20 milliseconds later. Your V1, that first bit of your visual cortex, responds about 40 to 50 milliseconds after the picture appears. From there every step along Highway What adds about 10 more milliseconds: a response in V2 10 milliseconds after V1; in V4 about 10 milliseconds after V2. Highway Do is faster, area MT's neurons changing their spikes a mere 10 milliseconds after that first detectable change in V1.[1]

Seems sensible, no? Highway Do is fast, a reactive route through the cortex for working out where and how something is moving, to give you immediate options for touching and grabbing, or ducking and diving. Highway What is slower, a deliberative route through the cortex for working out what something is, to let you know if it's edible, punchable, or about to castigate you for being late for work. Slower, but your fusiform face area will still light up less than 100 milliseconds after seeing a familiar face.

Doing more than just responding to the outside world doesn't take the brain much longer.[2] Here's a challenge for you: I'll flash you a picture for a mere 30 milliseconds while you hold down a button; your job is to let go of the button only if the picture contains an animal. From that merest glimpse, your brain has to use the spikes pouring from the retina about where is dark and where is light and what angle the dark and light fall at, use those spikes to not only reconstruct the basics of what is in the picture, but then compare that reconstruction to some stored memories of what animals look like. And then decide whether any of those animal-like features are in the picture. Sounds tough. Yet in all probability you would get at least 90 percent correct. Because your brain can work this out really fast: it takes just 150 milliseconds from the picture first appearing until activity at the end of Highway What (in your prefrontal cortex) differs between pictures with and without animals.[3] And if that sounds implausibly fast, then consider this: Terry Stanford and friends showed that just 30 milliseconds is enough for a new bit of visual information to impact a decision.[4]

Of course, the more processing the brain has to do, the slower the response. Once your brain has decided whether the picture contains an animal or not, there's more work to be done. You have to select the correct response—release the button or not—and then execute that response. You'll likely average around 450 milliseconds between the picture appearing and letting go of the button. Which as your brain already seemed to know whether there was an animal as early as 150 milliseconds, suggests that sorting out the correct response took a long time.

Or we can make your brain slower by making it do math. To you this might seem blindingly obvious. Nonetheless, Stanislas Dehaene set out to test quite how slow.[5] He asked his volunteers a simple question: is this number larger or smaller than five? They averaged about 400 milliseconds between their first exposure to the number and giving the answer by pressing a button. Their brains took less than half a second to comprehend what the number was, compare it to five, and then respond by pressing the correct button. And could be even faster.

The clever bit of Dehaene's experiment was that he manipulated all three parts—comprehending, comparing, responding—to find out which was the bottleneck. Responses were faster if the number was shown not spoken, suggesting the brain comprehends writing faster than speaking. Responses were faster for numbers farther from five, suggesting the number line is literally real. And responses were faster for button presses with the right hand than the left hand: as all volunteers were right-handed, this suggests that the dominant control of that hand by their left motor cortex (the lateralization of chapter 4) plays out as faster processing speed. So the fastest response of 375 milliseconds came from seeing a number far from five that needed the right-hand button as its response; the slowest response of 435 milliseconds from hearing a number close to five that needed the left-hand button. But these differences in processing—in comprehension, in comparing, in responding—shifted the brain's overall responses by just a few tens of milliseconds.

We can make the brain slower still by not giving it enough information. Those movies of dots randomly moving about pop up again here. Recall that the task is to decide the dominant direction that the dots are moving in and indicate that decision by looking at a light in that direction—to a light on the left, or a light on the right. We can control how hard the task is by changing the fraction of dots moving in the same direction. And the fewer the dots moving in the same direction, the harder the task, the slower people, monkeys, and rodents are to make a decision.[6] If half of the dots are all moving in the same direction, then it takes about 400 milliseconds to make a decision, and almost all decisions are correct. But drop that to just 3 percent moving in the same direction and then it takes about twice as long to make a decision, and still subjects make many, many errors. Worse, if we play the cruel trick of making no dots move consistently in the same direction, so there is no right answer, people will stare at the display for a second or more, even when instructed to make a decision as fast as possible. When gathering sparse, difficult information the brain slows right down. Slow is still relative—it's a mere second, after all.

Your decision to pilfer the cookie in your hour of need combines all the above, and more. Spikes combining all those patches of light and dark and angles into a crumbly cookie, within the tray of a box with lid open, "Cookies" scrawled on it, sat atop the bland brown desk. Spikes for recognizing that oaty-brown bobbly half-moon dotted with dark chunks as a cookie, an item of food, an answer to the pressing problem of rousing yourself before the imminent all-hands meeting. Spikes for remembering where your coworkers were just a split second ago, who is with whom, who is looking where. Spikes for accumulating those memories and new information as evidence for whether to take that cookie or not. Spikes for reaching, reaching to touch your fingertip to its crumbly edge. Even given a generous allowance for the lethargy of your postlunch addled brain, that's about 300 milliseconds to assemble the what and where of the cookie, another 1.5 seconds to recall everything and make a decision, and another 300 milliseconds to lean and reach for joy. All that in just 2.1 seconds.

On these timescales, spikes are ponderous. The physical process of making and sending a spike places hard lower bounds on how many spikes can be transmitted, received, and made anew within a second or two. Evoking a voltage blip is fast, but not infinitely so: molecules diffuse, ions flow, the voltage rises and falls. Collating blips to reach a tipping point can be fast, but not infinitely so. Making the spike at the tipping point is fast, but not infinitely so. Sending a spike along an axon is fast, but not infinitely so. Each step of making and sending a spike adds more time, more delay to the processing of what is happening in the world. Even if each spike arriving at a neuron evoked a voltage blip big enough to push the neuron to its tipping point, there'd still be a delay of at least 10 milliseconds from the spike arriving at the gap to the new spike arriving at its destination—and longer still if the axon is slow or long or both. How then can you know what and where a cookie is in less than 300 milliseconds? How then can your brain overcome the speed limits of spikes?

There are two solutions. The first solution is well known: the brain computes in parallel.

ROADS NOT TAKEN

This first solution to the speed limit problem lies in the roads not taken. Our journey was a single path, a single, serial chain of neurons, stretching from the retina, through much of the cortex, and down to the spine. And even then we had to clone ourselves, one of us down Highway What, the other down Highway Do, to keep track of the division of labor needed to solve the seemingly simple task of picking up a cookie. But there were so many roads we could have gone down, journeys that were happening in parallel to ours.

We could see this parallelism all around us. Every time we clung to a cloned spike zipping along its axon to a gap, we jumped across it to descend a neuron that was about to send a spike. We could have taken the leap at any of the gaps that axon made. How many roads could we have walked down?

As we now know, we got lucky. A single axon makes many gaps. But failure rates are high, rendering the spike ineffective at most gaps. And even if effective, the neuron on the other side is most likely dark, not receiving enough input to make a spike anytime soon. A spike's journey is a perilous one, prone to disaster.

Let's run the numbers to see quite how perilous.[7] We discovered a pyramidal neuron makes about 7,500 excitatory contacts with other pyramidal neurons. And that these contacts have a failure rate of about 75 percent. Given how many jumps we made between neurons in those 2.1 seconds, we had to be on our way again in 10 milliseconds or less for the new spike to have any impact on you taking the cookie. The probability that each of those 7,500 neurons will send a spike in the next 10 milliseconds is (very roughly) 1 percent. So when we started following a spike from a pyramidal neuron somewhere in the cortex, there were just 19 other neurons we could have reached to continue our journey in good time. Nineteen neurons at which the arriving spike would not have failed, and would then send their own spike quickly enough. Or to put it another way, we had to bulls-eye one of just 19 options out of 7,500 to carry on our journey at each leap between neurons. Threading our way through the whole brain in just 2.1 seconds, we got lucky indeed.

Ah, but even with this tiny number, the number of possible paths we could have taken explodes with just a few jumps. For those 19 neurons in turn will also be able to reach that same number. So in two jumps, there were 361 paths we could have gone down. In three jumps, 6,859 paths. Even starting from a single neuron, and even with the gambler's luck needed to find the scant few options onward, within just a few leaps the number of potential paths not taken, of parallel routes fanning out from our starting point, explodes exponentially. And each of these paths is potentially computing something different in parallel.

There are so many more roads than this parochial parallelism. Each region of the brain sends spikes in parallel. Indeed, even just departing from the retina, the retinal ganglion cells tile the visual world—nearby neurons responding to nearby things; neurons at the bottom of retina responding to things at the top of world, and vice versa. The pixels of the cookie scene were dealt with in parallel within each channel: some ganglion cells handling the rest of the detritus strewn across the edge of your desk and spilling onto the abutting one; some a glimpse of a wall twixt Graham, Graham's tie, and the (broken) photocopier; yet others the motivational poster on the room divider ahead of you proclaiming "Customer Service is not a department, it's a state of mind." We followed the spikes from but one location in space, from a pixel of crumbly cookie edge, while all around us the spikes of other ganglion cells were conveying the rest of the world in parallel.

And then each of those pixels is handled by at least thirty different types of ganglion cell crammed together. That each convey different information, about the onset or offset of light, of how fast light is changing, or its direction, or combinations thereof. Each pixel in space computed in parallel, and within each pixel more than thirty separate streams of information in parallel, all carried by spikes. And we followed just one.

We saw this parallelism at the end of our journey too. Neurons in the motor cortex have to send information in parallel. Some sent spikes to change your posture; some sent spikes to move your shoulder; some sent spikes to extend your arm.

Beyond this areal parallelism, the whole brain is one vast parallel computing monster. There was so much of the brain we could not visit on our journey. Spikes from the retina sent directly to the brain stem, to make your eyes move rapidly to look at any new, important, and potentially hungry thing that just popped into view. Spikes sent to and around the hippocampus, recalling memories of similar incidents of premeeting emergency blood sugar boosting, creating memories of this cookie incident, and keeping track of where you are in the office. Spikes sent to and around the amygdala, ready to learn from any bad outcome of the cookie-nabbing escapade. Spikes to the thalamus, the cerebellum, the substantia innominata, the hypothalamus—to everywhere. Spikes sent all over the brain to compute different things, offer different solutions, all in parallel.

The brain is parallel at all scales, from the paths emanating out from a single neuron to the simultaneous routes through swaths of the brain. Such parallel processing solves one part of the speed limit problem, of how to do everything at once, by dividing up the world and computing each bit of it at the same time. But it doesn't solve the main speed limit problem: that each individual thing being computed—the pixels of light and dark turned into cookies into decision into movement—is still happening in sequence.

In less than a second, most neurons do not have time to make a spike. And the few that do can send at most a handful. And even then, the last few of that handful will arrive at their target neurons after those neurons have already sent their own spikes. How then do we get spikes from the eye to the front of the cortex in under 150 milliseconds? We need another solution.

SPONTANEITY FOR SURVIVAL

Imagine if the body had to wait for spikes from the retina to make the legion of spikes in the first vision area of cortex, to in turn make the legion of spikes in the second vision area of cortex, and on and on all the way to the spine. This would take tens of seconds, minutes even.

Bodies don't have time to hang around waiting for legions of spikes to be created from scratch at each step. Hanging around gets you eaten. How then are spikes turning sensation into verbs—lifting, reaching, moving, deciding—fast enough for us to survive?

The solution, I argue, is the spontaneous spikes. They are the brain's solution to the deep problem of survival. The body doesn't have to wait for them to be created, as are they are already there. And as they are already there, they can be put to work.

Spontaneous spikes can be put to work to make neurons respond faster to the outside world. Left to its own devices, unperturbed by input, a pyramidal neuron's voltage will default to a value way below its tipping point. Starting from this state of repose is why it takes a few hundred input spikes to make one output spike from that one neuron (chapter 3). And that's not even including the vampiric effects of GABA, quashing attempts to drive the neuron to the tipping point. Like you, a reposed neuron is sluggish when roused and prone to falling back asleep without the constant blaring of an alarm.

But what if the neuron's voltage was already close to its tipping point? Then the neuron needs to receive just a handful of extra spikes to make a new one. Can send a spike almost instantly. If those handfuls of extra spikes are evoked by something in the world, the neuron can respond to the world almost instantly. And how can a neuron be near its tipping point when spikes evoked by the world arrive? Exactly: spontaneous spikes, causing voltage blips in many neurons, driving up their voltage.

The swirl of spontaneous spikes around a big circuit of neurons can ensure that there are always some neurons close to their tipping point. So whenever new spikes carrying messages about the outside world arrive at that circuit, those neurons can react to them almost instantly, sending their own spikes in response, carrying the message onward.

While a single neuron can be sluggish, a group of neurons is always ready to respond—the legion in action again.[8] And with these primed neurons in each region of cortex, in V1, V2, V4, MT, prefrontal cortex, the lot, each step of our journey could be taken within a few milliseconds.

In this view, spontaneous spikes are there to assist spikes evoked by the outside world. It clearly works but raises a host of niggling issues. For one, the whole thing is happening by chance: at the precise moment that new information arrives from world, which neurons are being moved to their tipping point by spontaneous spikes is likely random. Perhaps it doesn't matter which neurons are randomly moving to their tipping point, if all the brain cares about is the information carried by legions of spikes (chapter 7). But even so, how can the brain tell the difference between spikes that are for making fast reactions in other neurons and spikes that mean something about the outside world?

The most profound issue is that putting spontaneous spikes to work as mere helpers for evoked spikes is a strange use of the huge amounts of energy needed to sustain the spontaneous activity. Indeed, so huge an amount of energy that spikes evoked by the world or by doing something in the world barely draw any more energy than the ongoing spontaneous activity of the brain.[9] That they use so much energy suggests the brain has a much better way of putting spontaneous spikes to work.

We think we know what that is: spontaneous spikes are already carrying most of the information the brain needs to act. They are prediction.

SPONTANEITY FOR PREDICTION

Prediction is crucial to much of what we do. Our built-up knowledge from prior experience guides our future behavior. That guidance is prediction: based on what happened before, this is what's most likely to happen next. Our brain turns experience into prediction. And this plays out over every timescale of behavior, and from the simple to the complex.

Predicting the Visible World

There's something you learned from experience that took so long you didn't even realize you learned it: seeing. The development of your visual system takes a long time. Your eyes opened in the womb, but there was nothing to see. When born, your brain didn't know what the world looked like. It didn't know the statistics of the world. Of how many edges, corners, and curves there are. Of where they tend to be—the consistent edges of horizons and tree trunks and houses, the consistent corners of paper and dice and windows, the constant curves of the moon and footballs and pies. Of how those edges, corners, and curves tend to relate to each other, to form trees and houses and footballs. And of how they tend to move, in graceful arcs and smooth trajectories, no sudden vanishing, reversing, or plummeting.

These statistics of the visible world are all learned by experience. Raise someone in a world with no vertical lines, and they will not be able to see a vertical object placed before them.[10] Raise someone with one eye closed, and when reopened that eye will see nothing.[11] In both cases, the neurons in visual cortex have not been able to learn the statistics of the world—deprived of the experience of vertical lines, there are no neurons tuned to vertical lines; deprived of the experience of one eye, there are no neurons tuned to the view from that eye. Neurons learn about edges, corners, and curves through experience. The set of neurons you have in your visual cortex exactly reflects the statistics of the visible world you grew up in.

Which means your brain can turn this lifetime of experience into prediction. If the visible world tends to have a consistent set of statistics, then we don't need to analyze each scene anew, for most of what is in the world right now your brain can predict, because your neurons have experienced it before. Predictions made by those experienced neurons firing spontaneous spikes. Those spontaneous spikes are predicting a specific collection of edges, corners, and curves in the world right now. Predicting their imminent movements; like the fact that Sarah moving away from you right now will in a moment be slightly farther away in the direction she's already traveling, as teleportation isn't a thing.

Predicting more complex, nuanced features, the colors, textures, and objects in the world.

These predictive spontaneous spikes solve the speed limit problem. We don't need to get spikes from the retina to alone create the first wave of spikes in V_1, to in turn recruit those in V_2, then V_4 and on and on. Because the spontaneous spikes are already there in each of those regions, are already predicting most of visible world.

How, exactly, does this prediction occur? One clever theory is that visual cortices partly solve the speed limit problem by predicting what information should be coming from the eye. As we ascended Highways What and Do, the neurons at each jump along the way sent spikes in response to increasingly complex aspects of the visible world. In this prediction theory, the spikes we traced on the way up Highway What are signaling what is most likely to be out there in the world, the best guesses at the moment: spikes from V_1 signaling the most probable set of simple features; spikes from V_2 the most probable conjunction of features, like long edges and corners; spikes from V_4 the most probable collection of those conjunctions and colors; those from the temporal lobe the most probable objects. But the key to this theory is what happens going the other way. Remember those spikes that whizzed back past us at every step, spikes that were already there, racing back down Highway What to whence we'd came? They are doing the predicting.

Working backward from temporal cortex, those descending spikes are predicting the features that should be in the world if each guess is true. And at each step back down the Highway, they predict by deconstruction: if this object is out there (temporal lobe), then it should have this set of complex features (in V_4); if these complex features (V_4), then this set of corners and curves and long lines (in V_2); if this set of corner and curves and long lines (V_2), then this set of simple edges arranged just so in space (in V_1). So the spikes descending Highway What ultimately predict what information (edges, where they are, their angles, and their location) should be arriving in V_1 from the eye.

And the input from the eye simply adjusts what is wrong with the predictions. As most of the predictions will be correct, because our visual system spent many years learning what is there in the world, there won't be much to adjust. So most of the spontaneous spikes in your visual cortices will be telling the rest of your brain an accurate version of what is there in the world, before your eye has even "seen" it.

That's the theory, anyway.[12] Bayesian hierarchical inference, to the cognoscenti; bootstrapping, if you prefer; educated guessing to the rest of us. Whatever we call it, it means seeing can be so fast you don't notice it happening; the information is already there and adjusted by the input.

Predicting What Will Be Useful

I'd argue the brain putting spontaneous spikes to work for prediction is way beyond just seeing, or indeed any of our senses. Prediction is everywhere.

Brains predict the outcome of decisions. To illustrate, let's play a simple game. You have to choose one of two cards I present. One card has a high probability of winning you a reward, the other low. (The reward is some chocolate milk.) Naturally, you'll want to figure out which of the two cards has a high probability, as who doesn't like chocolate milk? If I repeatedly present the pair of cards, you'll quickly figure out which card is best. You'll come to expect that card to give the highest payout, to predict the likely outcome of choosing that card. You'll then use that prediction to guide your behavior; when presented with that card, you'll predict goodness from choosing it, and so will choose it. Your brain can create predictions quickly from a few examples of simple events.

So we should be able to find neurons whose spontaneous activity predicts good things are about to happen. Michael Platt and Paul Glimcher reported just such a set of neurons in the back end of the parietal cortex.[13] They asked their monkeys to do a simple task: stare at a central spot then, when signaled, choose to look at the light above or at the light below. It took just a few trials for the monkeys to learn which of the two lights was paying out the most juice. And for their neurons to predict which light was the most valuable. For in the back end of parietal cortex are a collection of neurons that fire just before the eyes move. One group fired a lot of spikes just before the eyes moved up, another group fired a lot just before the eyes moved down. The spontaneous spikes of these eye-movement neurons changed to predict the value of the light. If moving the eyes up brought more juice, so the upward-movement neurons fired more spontaneous spikes before the signal to choose. If moving the eyes down brought more juice, so the downward-movement neurons fired more spontaneous spikes before the signal to choose. And the greater the value of the best light—the more juice was gained over the other choice—the more spontaneous spikes were fired by the neurons that would move the eyes to that light. The spontaneous spikes were predicting the value of the action represented by that neuron.

We see spontaneous activity predicting decisions all over the brain. During that pesky randomly moving dots task, before the dots even appear, eye-movement neurons in that same region of parietal cortex are spitting out spontaneous spikes. And the more they send, the more likely the monkey will decide the dots are moving in those neurons' preferred direction.[14] In a monkey's V1, the greater and more correlated the spontaneous activity, the more likely that monkey will detect that a bit of a picture has been rotated to look weird.[15] And in people trying to decide whether the picture they're looking at is two faces or a vase, the more spontaneous activity in their fusiform face area before seeing the picture, the more likely they are to decide it's a face.[16]

Theories of how the brain does decision-making propose all these effects of spontaneous activity on future decisions are because that activity encodes prior information. That a particular neuron's spontaneous spikes before a decision mean: given this body's previous experience, this is my current prediction of how likely my option is to be correct or valuable. (Whether "my option" be which light to choose, which direction the dots were moving in, that we're looking at a face etc. etc.) And all such neurons send their predictions before any information comes from the outside world, so those predictions can be used as the starting point to make the upcoming decision as fast and as accurate as possible.[17]

Holding things in your memory buffer is also prediction. There is a strictly limited capacity for what we can keep in that working memory. So for something happening in the world to gain entry to this buffer means it must be likely worth remembering. Placing an event in that buffer is then a prediction that it will be useful to know in the immediate future.

We can see this prediction in the two ways of gaining entry to the memory buffer. Of its own accord, your brain tends to lodge things there that are new or odd or surprising. It disregards routine things. Like the action of locking the door when you leave home for the day; within seconds of locking the door, you likely have no memory of having done so. For routine, mundane things are highly predictable, habitual, with a low likelihood of something about to happen that will need a memory of that routine, mundane thing. But the appearance of new, odd, or surprising events are predictive that you may need to take some action about that event in the future.

The second way is that our memory buffer also retains memories of events that have proved useful through experience. In the excruciatingly dull tasks we set ourselves and other animals to probe the brain's memory buffer, we establish that an event needs to be stored in that buffer by repeating it over and over again. We might flash a light in a particular location, or vibrate a strip of metal at a particular frequency. Such events predict future reward, but only if we use that remembered information to guide action—like looking at where the light was, or reporting whether a second vibration was higher or lower than the first. So we and other animals learn to pay attention to these things, to deliberately lodge them in our memory buffers, as they predict the future action we ought to take.

In everyday life, we can use the same idea of deliberate concentration to lodge an apparently routine event in our memory buffer. By concentrating on it, giving it our attention, we are signaling that it will be useful in the future. From long experience I know after leaving the house the first thing my wife says to me as I get in the car will be "did you lock the door?" And if I can't look her in the eye and say "yes" with absolute certainty, I know I'm inevitably heading back to the front door to check it. So from such long experience I have learned to now concentrate my attention on locking the front door before I get in the car, lodging that action in my memory buffer, predicting that I will need that memory in but a few seconds' time. By my calculations, over many years of marriage that simple act has gained me a few hours of my life back not spent reluctantly trudging to and from the front door.

However those memories are lodged, we already know from our journey through your brain that they are all maintained by the spontaneous activity of neurons in the prefrontal cortex. Long after the evoking event, these neurons are sending spikes to each other, maintaining each other's activity, predicting that this event will be useful imminently. To guide imminent decisions, to guide future actions.

Actions like moving your body. And on our journey we've already seen the prediction of movement. In the motor regions of cortex we ran into a barrage of spontaneous activity, neurons firing long before movement started. Spontaneous activity that prepared movement, that predicted movement, readying the legion of neurons that directly control the movement to respond. In humans, the level of spontaneous activity across the motor regions of cortex (as measured by fMRI) seems to dictate how strongly we press a button.[18] In monkeys about to move their arm, the spiking of neurons in premotor cortex is proportional to the probability that the monkey will shortly move the arm in those neurons' preferred direction.[19] Spontaneous spikes before movement are also seemingly predicting what that movement will be.

Spontaneous activity as prediction can solve the speed limit problem everywhere. You don't have to wait for making a decision from scratch each time, as spikes are already preempting the likely decision. You don't have to regather sensory information anew, as spikes are already maintaining what could be imminently useful. You don't have to create each movement from scratch, as spontaneous spikes are preempting the likely next move. Your spontaneous spikes let you respond faster, better, more—no hanging around, no being eaten.

The Domination of Spontaneous Spikes

We have clever theories about how the visual bits of cortex could use spontaneous spikes to make predictions about the visible world. And I've proposed ideas for where prediction resides outside the sensory realm. These theories and ideas may or may not turn out to be true. But if the spontaneous spikes of neurons in your cortex really are predicting the world, then, no matter how they are doing it, a simple fact should hold true: the spontaneous activity from neurons will consistently look like activity evoked by something happening in the outside world.[20] Because if the evoked activity is often unique, and looks nothing like the spontaneous activity, then spontaneous activity cannot be making predictions.

It's easiest to test this idea in V1. We have a really good idea of how neurons there respond to the visible world—see, for example, most of chapter 3—so we know what the spontaneous activity in V1 should look like too. And we have a wealth of evidence that spontaneous spikes in an adult's V1 look just like evoked spikes.[21] Grinvald and his colleagues provided much of this evidence in a sequence of three crucial papers.[22] In 1996, they showed we can predict the activity evoked in V1 by a picture using the spontaneous activity just before the picture appeared. Better yet, the evoked and spontaneous activity are exactly as similar as the spontaneous activity is to itself a little while before the picture appeared: there is nothing special or unique about the evoked activity. In 1999, they showed that what makes a single neuron in V1 fire is the same pattern of activity appearing in the surrounding neurons, regardless of whether that pattern is evoked or appeared spontaneously. The legion in full effect there. And in 2003, they showed that groups of neurons that preferred the same angles of edges in space are spontaneously active together, just like they are when presented with the world to look at. And there's more. Fiser and colleagues showed us that the correlations between spikes in V1 are the same whether the spikes are the spontaneous activity of eyes in darkness or evoked by watching a movie.[23]

We've seen this matchup between spontaneous and evoked activity elsewhere in the cortex too. In the first bits of a rat's cortex dedicated to sound or touch, the sequence in which a group of neurons fire is the same whether the rat is listening to sounds, is asleep, or is anesthetized.[24] And my lab has shown that in the prefrontal cortex of those rats running around in a maze, the pattern of spikes across the legion of neurons while they explored the maze reappeared in sleep.[25] So often in fact, that almost every millisecond of running in the maze was accompanied by the patterns of spikes that recurred in the spontaneous activity of sleep. Truly, there is nothing unique about the spikes evoked by events in the outside world.

We've even seen the spontaneous activity in the visual cortices changing during development, as the brain experiences and internalizes the statistics of the visible world. In a beautiful experiment, József Fiser and friends, led by Pietro Berkes, traced the development of V1 in ferrets by recording at different stages of maturity the patterns of neurons firing in response to a movie of the natural world and in total darkness.[26] Infant ferrets had vastly different patterns of spikes between movies and darkness. But these converged over development, so that in fully grown adult ferrets, similar neurons fired at the same time whether they were looking at movies of the natural world or sat in pitch darkness. (The "natural" movie was the trailer for *The Matrix*. By "natural" they clearly don't mean trees and flowers and bees. In fact, it's hard to think of many films more unnatural. What they mean is a series of images that contain the usual statistics of the world: lots of edges and curves and corners, in the usual relationships, moving in the usual way. Some of which happen to be Keanu Reeves in a trench coat.)

And how do we know this convergence is from learning the statistics of the visible world, as I claimed above? Because this convergence was specific to natural images. When viewing unnatural images made up of just lots of straight, parallel lines, the evoked patterns of spikes remained distinctly different from the spontaneous patterns happening in darkness. Which all together suggests the development of predictions in the visual cortex. That the predictions of spontaneous activity make large errors in infants with just-opened eyes because the statistics of the visible world are unknown. That the predictions get better as the world is experienced over time, the errors get smaller, and so the difference between the spontaneous and evoked activity gets smaller. And so by the time we reach the mature visual cortex, there is very little difference between the spontaneous and naturally evoked activity, as most of the natural world can be predicted. But the errors in predicting unnatural images never go away, as they are not part of the ferret's experience of the world.

THE ETERNAL CYCLE

I'm proffering here a simple but rather radical model of how your brain works. Spikes coming in from the outside world adjust the spontaneous activity, and these adjustments are the messages they carry. Like molding Play-Doh into a snowman, then a cookie, then a tree—all the same stuff, but adjusted into different meanings by the prods coming from your fingers. The adjustments are errors in the predictions: of what was sensed, of what was enacted, of what was the consequence.

This is why in mature brains the distribution of how many spikes each neuron sends barely changes between spontaneous and evoked activity,[27] nor as we've just seen does the distribution of the patterns of spikes in V1 and prefrontal cortex. Most spikes are born of ongoing, internal, spontaneous activity, and are merely prodded into new forms by input. Which means that information is not coded in the pattern or timing or number of spikes evoked by the outside world; it is coded by changes in the ongoing, spontaneous spikes.

If so crucial, where did they come from? I'd argue spontaneous spikes are an inevitable consequence of evolution wiring up a big bag of neurons into a brain. Neurons are essentially the same in everything with neurons: humans and leopards, snakes and frogs, ants and worms, zebrafish and squid. Which is one reason why I could write this book and be able to describe how your brain works despite us not being able to record spikes from human brains. And which means neurons must be evolutionarily ancient, originating in a common ancestor to every living thing with a nervous system. Our best guess is they appeared somewhere between 635 and 540 million years ago.[28] Either they evolved to coordinate rapid responses to things happening in the outside world, like to avoid being eaten,[29] or, as evolution moved from single-celled organisms to multicellular life,[30] they evolved to solve the problem of coordinating between those newfangled cells to make sure the cells doing moving and cells doing feeding were doing the right thing at the right time in the right order.

Neurons clearly proved their worth for survival. They are everywhere, after all. So more were added. And as soon as we couple multiple neurons into a network, we get multiple feedback loops; we get the potential for spontaneous spikes from the network alone. This is especially true once the number of neurons is so large that each neuron and its connections cannot be individually specified by genes. Then genes can specify types of neuron, where they appear in the nervous system, and which types of neuron to wire to. But the details of the wiring is left to chance. A feedback network of neurons is then inevitable. With that, spontaneous spikes appeared as a by-product of simply making a network of neurons.

I'm suggesting evolution co-opted them to do useful work, to solve the speed limit problem of brains. A problem created by adding more neurons. For as soon as there appear many more neurons between those taking sensory inputs and those directly making movement happen, there is much more processing to do, more delays to overcome, more possible ways of mapping input to movement. Evolution could then co-opt spontaneous spikes to prime responses, or anticipate needs. And from this evolve the complexity of the endlessly predicting human brain.

Our journey showed the deep complexity of the apparently mundane. An act over the briefest of moments invoked a staggeringly complex cascade of spikes. Billions of spikes from billions of neurons along billions of axons, leaping billions of gaps. Each spike cloned at each branch in the axon, multiplying endlessly, each clone racing along the cable, trying to trigger the cascade of molecules across the gaps to the tree of the next neuron. If it succeeded, and we know now it often does not, the blip of voltage evoked in the tree joins thousands of others in that moment, each blip sliding down the torturous tree, some exciting the neuron to its tipping point, some inhibiting those blips, crushing them. Most of those thousands of others were spontaneous spikes, dominating all. The constant back-and-forth of spikes across the brain, swirling around the layers of cortex, between the regions, up and down through the rest. They are always there, and they vastly outnumber the spikes evoked by the outside world.

Brains work on many slower timescales than spikes.[31] Neuromodulators like dopamine and serotonin change over minutes. Changes to synaptic strengths to create immediate learning takes hours. Long-lasting changes to master complex skills takes days and weeks. Developing a brain takes years; in humans well into our early twenties. All alter how the moment-to-moment messages of spikes are sent. But none of them directly make things happen right now. Their effect is only felt through how they change spikes.

Moment to moment, the spontaneous spikes are crucial for us to operate in this world. In the sensory parts of the cortex, the spontaneous activity predicts what the next sensory input will be, ensuring that the brain can track the world quickly. Spontaneous activity in motor cortex and other regions prime the body to move, in the null space, ensuring the command for the next movement is just a few spikes away from happening. In the deep recess of the prefrontal cortex, spontaneous spikes hold memories of information needed to act, and hold predictions of the consequences of those actions, so decisions can be made quickly. But for humans at least, we know spontaneous spikes are far more.

Moment to moment, the spontaneous spikes are You. They are your daydreams and idle thoughts, contemplation and planning, memories and musings. Wondering what you'd look like with a vibrant purple streak in your hair. That sudden inspiration to sack off cooking and get fish and chips on the way home tonight. Conjuring that image of mastering guitar in a few years' time. A reverie of squidging toes into the wet sand as the sea gently laps your ankles like a slow dog. Your rich internal life is the sending and receiving of spontaneous spikes across your brain.

The most important journey for the spike, then, is not from input to output—it is the eternal cycle, looping forever within the brain.

CODA

The Future of Spikes

ALL THE SPIKES

Our spike's journey through your brain will soon be different. I could only interpret what we saw through things we already know, knowledge painstakingly gained over the past hundred years or so. After more time has passed, more data gathered, more knowledge mined, we will know more of the journey, understand better the things we saw. What might we know?

Such fortune telling ranges widely in difficulty. Easiest is to predict the kinds of data about spikes we will get in the future, for we can take current trends in technology, survey current proofs of concept, and extrapolate out. Harder is to predict what insights those data will contain. Hardest is to predict what they will mean. But what we can predict are the new directions we want to explore. And what we want to explore is everything that is missing entirely from this book because we know nothing about them: spikes that underlie disorders of the brain, and spikes that underlie human thought processes. In order to be able to get at these spikes, we may first need that extrapolation of technology.

Driving this Golden Age of systems neuroscience is the white heat of a technological arms race, a race to record as many spikes from as many neurons at the same time as possible. Indeed, neuroscience has its own equivalent of Moore's law, with the number of simultaneously recorded neurons doubling every few years. In 2011, the number of recorded neurons was predicted to double every 7.4 years; now, in early 2020, the predicted doubling of neurons is every 6.4 years.[1] There's no sign of a slowdown in our ability to grab more neurons.

The "Neuropixels" probe beautifully illustrates the kind of leaps that make up this doubling. It arrived in 2017, a slender silicon thread for inserting deep into an animal's brain, its 10 millimeter length packed so densely with contact sites that each probe can record the individual spikes emanating from up to two hundred neurons.[2] Implant a few of them at the same time, and one quickly gets many hundreds—closing in on a thousand.[3] These neurons are widely dispersed throughout the brain. For these probes are long: a mouse's cortex is less than a millimeter thick,[4] so a 10 millimeter shaft passes deep into its brain, collecting neurons from many regions, regions hitherto never recorded together at the same time.

Which makes for a simple yet deep prediction about our understanding of spikes in the near future: it will be upended again. By happening to record from many brain regions at the same time, we will upend many nice theories about how a bit of brain X is responsible for doing thing Y. When we look at spikes in many brain regions at the same time, we will find many regions seem to be involved in the same thing—in deciding, moving, remembering, perceiving—and many of those regions will, crucially, not be in the cortex. Indeed, we may be on the cusp of a new, less cortex-centric view of the brain.

More concretely, we can make two pretty firm predictions about the types of data we will get in the near future. The first is simply more spikes, from more neurons, than we could possibly imagine just ten years ago. The present capacity for imaging calcium in neurons hints at how many more spikes we can get. Its upper limits are the imaging of about ten thousand simultaneous neurons in mammals, and tens of thousands in baby zebrafish.[5] (Why baby zebrafish? Because they have translucent heads! The glowing chemical in their brain can just be videoed from the outside.) Calcium imaging does not directly record spikes; it records the slower changes in calcium inside the neuron's body that are caused by spikes, which is a useful proxy, but there is no clean one-to-one correspondence between spikes and changes in calcium.[6]

Nonetheless, the development of calcium imaging means we have already developed all the kit we need—the microscopes, the rigs, the analysis software—to image many thousands of individual neurons. So all we'd need do is replace the chemical that glows in proportion to calcium with a chemical that glows in proportion to the neuron's voltage, and we have in principle the capacity to image the spikes from thousands of neurons.

And we (almost) can.

It's called voltage imaging. Directly videoing the glow of chemicals that responds to changes in voltage. In truth, it's been around for a few decades.[7] But up until now, we could only use it to look at the spikes of single neurons in simple invertebrates, in leeches and sea slugs, because they have giant neurons with few spikes.[8] Simply because imaging voltage is doubly handicapped compared to calcium imaging; voltage-imaging is trying to record much faster things—spikes—with far, far less of the chemical available to glow in response. Instead of filling the neuron's body, like the chemicals for calcium imaging, the voltage-sensitive chemicals can only live in the neuron's skin because that's where the voltage is changing, as we learned right at the start of chapter 2. Roughly, the amount of the calcium-sensitive chemical is proportional to the volume of the neuron's body, but the voltage-sensitive chemical is only proportional to its surface area, giving a hell of a lot less voltage-sensitive chemical to detect. So only really giant neurons—with a vast surface area—could contain enough of the voltage-sensitive chemical to have their changes in voltage detected. And only in invertebrates do we find these giant neurons, with bodies tens of micrometers across. Only in these creatures could we use this magic of videoing voltage directly.

That's all about to change. Multiple breakthroughs in new types of voltage-sensitive chemicals happened in 2019, making them glow much brighter and change much faster and last much longer. Imaging the voltage of many single neurons at the same time is at last working robustly in mammals.[9] Now all we need is scale, to move from handfuls as now, to tens, to hundreds, to thousands of neurons.

The best use of voltage imaging is yet to come, and this use is the second prediction about the type of data we will get in the future. Because we are imaging the neuron's voltage, we can in theory see more than just the spikes. We can see all the flickers of voltage in between each spike a neuron spits out. See all those blips created by its inputs. This needs ultrabright, ultrafast, ultrastable chemicals, and they're coming too. And with all those flickers at hand, we can see what caused the spike itself. We can trace the journey in detail from spike to blip to spike to blip to spike to blip and on and on . . .

If the exponential explosion in recording neurons runs its course, one day we will be able to record every single spike sent by every single neuron in the entire cortex of a mouse. Oh happy day! What marvels we will see, what things we shall learn! Except: imagining such a future brings into sharp relief the "so what?" question. So what do we do with these data?

For if we keep digging up more spikes, do we advance our understanding of the brain? Or do we instead simply fragment our understanding, driving deeper and deeper into the details, piling more observations and facts on each other, drowning in Big Data.[10]

We must be wary of the latter. For what we've perhaps learned most from this recent Golden Age is the yawning chasm between what we thought we knew about how brains work, and what we actually know. On our spike's journey, we've seen how digging deeper into spikes has taught us much that is new about how the brain works. About how a single neuron's dendrites, its tree, combines its input spikes in clever ways. About how synapses seem to fail on purpose. About the ubiquity of the dark neurons, and the mystery of their role in life. About the level at which the brain encodes information—groups of neurons, the legion—rather than the single neuron. About spontaneous activity being not noise, but having purpose, of overcoming the speed limits of the brain's machinery.

The most obvious chasm in our understanding is in all the things we did not meet on our journey from your eye to your hand. All the things of the mind I've not been able to tell you about, because we know so little of what spikes do to make them.

SPIKING ERRORS

If the passing of spikes between neurons is the basis of thought, word, and deed, then errors in the passing of spikes are what drive errors of thought, word, and deed. Some of these are fleeting, like the slurred speech of the inebriated, or the irritating clumsiness of the sleep-deprived new parent, valiantly struggling to close the nappy before another torrent of wee in the face.

Some of these errors are permanent. We call them brain disorders.

Some brain disorders are clearly errors in passing spikes. Epilepsy is a prime example. The seizures epileptics suffer can be either convulsive, where sudden muscle contractions shake the body uncontrollably, or nonconvulsive, as in absence seizures where the sufferer abruptly loses then regains consciousness. The capacity for epileptic seizures can be acquired via many routes—some are linked to specific genetic mutations, some from the playing out of genetic influences on development, and some from direct damage to the brain through strokes or tumors. But whatever the type and however it is acquired, the immediate cause of a seizure is an uncontrolled explosion of coordinated spikes across the brain, waves of neurons sending spikes in synchrony, especially those of the cortex and hippocampus. An explosion so violent in its synchrony, we can see it from electrodes placed only on the scalp. The waves of spikes drive the random muscle spasms in convulsive seizures and drive the loss of consciousness in absence seizures.

Many brain disorders are less obviously about errors in sending or receiving spikes. These are broadly divided into three classes: disorders of movement, of memory, and of thought. Study of these disorders concentrates on changes to aspects of brains that are not the sending of spikes. But the symptoms of the disorders all must ultimately be expressed by the way those changes in the brain in turn change spiking.

For disorders of movement, this is perhaps not so controversial an idea. Take Huntington's disease. The classic symptoms of Huntington's are the "chorea," the jerky, involuntary movements of the limbs. Huntington's disease is a rare case in that we know the specific single genetic mutation that causes it—a vanishingly small number of diseases, brain or otherwise, can be so linked to a single mutation. In Huntington's, this mutation is too many repeats of three letters of DNA—CAG—in a single gene, which in turn creates a mutated form of the protein encoded by that gene (the protein being called "huntingtin"—see what they did there?). That gene is mostly expressed in a group of neurons in the big, silent striatum, tucked just underneath the cortex. So those neurons get filled with a crap version of the protein, malfunction, and start to die off (why this only starts to happen when the mutation carriers reach their mid-thirties or later is not clear). As you now know, the striatum happens to be intimately involved in the control of movement, especially in making sure appropriate movements are happening. The death of many of its neurons means it can no longer send spikes that control appropriate movements. And this error creates the inappropriate, often violent, limb movements.

Disorders of memory are also errors in spiking. Alzheimer's disease causes profound loss of memory, alongside problems with cognition and changes in personality. Much work on Alzheimer's focuses on the accumulation around neurons of bits of proteins that shouldn't be there (beta amyloids), and errors in the folding of a protein inside neurons (tau tangles). These problems with proteins lead to malfunctioning neurons and malfunctioning connections between them, particularly in the cortex and hippocampus, neurons that ultimately die in droves, driving away memory in the process. For the recall of memories is the passing of spikes between neurons. As the errors in passing spikes accumulate with the increasing loss of neurons and of the connections between them, memory fades.

Disorders of thought are, too, disorders of spiking errors. Schizophrenia is a mélange of symptoms—taking in hypersensitive senses, distorted cognition, delusions, and hallucinations—whose root causes are unclear, and for which there are many theories. Hallucinations of sound, especially where the sufferer hears voices that do not exist, are common. Yet hearing is the passing of spikes between neurons, from those in the cochlear nucleus of the brain stem up through a chain of neurons to those in the bits of cortex specializing in sound. Hearing things that aren't there are then errors in the passing of these spikes, including the passing of spikes that shouldn't exist, that have occurred as though there was sound coming from the outside when none actually exists. And indeed when we scan the brains of patients suffering from sound hallucinations, their sound-specialized cortex is lit up just the same during hallucinations as during hearing real speech.[11]

Spikes give us a common language to talk about what in the brain goes wrong. Brain disorders have many root causes, whether they be mutations in a specific gene, a breakdown in the clearing of garbage proteins, a malfunctioning receptor for a particular chemical, a defect in fixing worn-out parts of cells, or prions—misfolded proteins—invading the nervous system and wreaking havoc. What they all have in common is their ultimate expression: all have a characteristic change to the passing of spikes between particular groups of neurons. And this change in the passing of spikes is what ultimately causes the symptoms of the disorder.

The problem is we rarely get to see what those changes are. Because we can't record spikes from humans. All the above is knowledge built from our understanding of the brains of animals. And mostly of spikes in normal movement, memory, and hearing, spikes from the brains of healthy animals. We know a little about changes to spikes in animal models of these disorders, animals with alterations to their brains that mimic some of the root causes of human disorders. But strangely little. Rarely are the current crop of Golden Age tools for recording many neurons at the same time used in these models.

A clear future for spikes is then to understand how they change in movement, memory, and thought disorders. In the near future, we will see a sharp uptick in neuroscientists using the newest recording technology in animal models to start mapping the changes to spikes across hundreds or thousands of neurons, across multiple brain areas at once. Hints of this uptick are appearing in recent studies recording tens or hundreds of neurons in animal models for Parkinson's disease and Fragile X syndrome (a rare form of autism).[12] And we would like to record these spikes in better animal models of these disorders, ones more closely matched to both the causes and the symptoms, for which we feel more confident in extrapolating to humans. A tall order indeed for many disorders of memory and thought, as it is hard to replicate things like dementia, depression, obsessive-compulsive disorder, and schizophrenia in an animal. Not least because the best models for us humans are those animals most closely related to us, the primates, and developing primate models of most memory and thought disorders would stringently test society's ethical tolerance for animal testing.

Weirdly, studying disorders by recording spikes in humans could end up being ethically more acceptable than developing new animal models. Routinely recording spikes in humans is an ethical nonstarter with current devices, because of the deeply invasive surgical procedures and sheer size of the devices. Such recordings are largely limited to the special cases of epilepsy (to find where the seizure-triggering activity starts) and Parkinson's disease (to locate the deep brain stimulation electrode in the right place). But a new wave of well-funded neural technology companies aims to change all that; they want to interface your brain's activity directly with a computer so are aiming to build implantable devices for the constant recording of spikes in healthy human brains.

Neuralink, for example, is creating neural lace, flexible electrodes that would flow through the brain, and with luck not trigger an immune response, allowing them to remain implanted for years.[13] Then there's "neural dust," the concept of nano-sized passive recording electrodes that can be read out using ultrasound.[14] At the time of writing in February 2020, these research programs are at best testing preliminary proof-of-concept devices in animals. But to reach their stated goals, at some point they will have to deal with the ethical hurdles of undergoing surgery to implant these things in a healthy human brain. Right now this too would seem a nonstarter; voluntary surgery on a healthy brain seems absurd given the risks involved, both those inherent in any major surgery from hemorrhage, infection, or accidents, and the additional non-negligible risk of permanent brain damage or stroke from fiddling about in the brain. But we know to never say never. After all, plastic surgery on perfectly healthy people is now routine. And there seems a better use of these devices, one perhaps more ethically acceptable: for understanding spikes and their changes in the array of movement, memory, and thought disorders unique to humans.

But if we ever could record the spikes in whatever human brain we'd like—your brain, my brain, the Dalai Lama's brain—then we know what else we'd go after: the spikes of subjective experience.

SUBJECTIVE SPIKES

Our journey through your brain traced the spikes of a single act, in but a brief moment of time, a handful of seconds at most. But there is a whole class of mental activity well known to you that I could not describe in spikes. Your plans and aspirations, your imagination, the mental imagery of situations mundane and absurd, your social interactions, your emotions, your attention and its control, your awareness, your inner monologue—your consciousness.

Why could I not describe these in spikes? Because—all together now—we can't record spikes from humans. When we do get the rare opportunity to record spikes in humans, we ask them to do simple tasks, like any other animal: choose between these two pictures for a reward; follow the moving dot with your eyes. All really dull stuff compared to the richness of our mental world. But for good reason. If we want to compare and contrast the coding and computation in human brains with those of other animals, we must use the same tasks.[15] And it's also just good science, keeping things simple, changing only what you want to measure. But no help for our understanding of spikes underlying thought, emotion, and the experience of being ourselves. Which creates a rather large explanatory gap: the passing of spikes between neurons is thought, word, and deed, yet of the above list of subjective experiences we know nothing about the underpinning spikes.

And into this explanatory gap can fall deep misunderstandings. We are left with at best vague generalities about the links between the brain's machinery, its neurons, and the aspects of mind with which we are intimately familiar. This allows a lot of weird things to manifest, like the rash of dubious concepts prefixed with "neuro-" to make them sound all science-y and smart—neuromarketing, neurolaw, neurocriticism.[16]

When we know literally nothing, absolutely nothing at all, not even a teeny-weeny iota of a quantum, about what signals neurons send to each other when you evince a deep loyalty to a brand of brown sugary soft drink, commit the crime of pilfering the last bit of Jamelia's milk from the work fridge, or offer a recondite interpretation of *Winnie the Pooh* as a deep metaphor on capitalism's destruction of childhood—for isn't the pursuit of honey merely an allegory for the lure of money? Sure, we can stick people in an fMRI scanner, and then report how changes in blood flow over a few seconds to an area that contains somewhere between millions and hundreds of millions of neurons correspond to the taste of two different popular, sticky brown drinks. But that tells us nothing about the spikes sent by the neurons. The things that do the work.

Of emotion we know a little. You may be familiar with the idea that the chunk of brain called the amygdala is the seat of "fear."[17] This is wrong. Evidence for the amygdala's role in fear is twofold. Rare people who lack a functioning amygdala seem to be fearless. And if we stop an animal's amygdala from functioning, it cannot learn that a sound or flashing light predicts a pinprick electric shock. The amygdala is the site of that learning, that some event in the world predicts an imminent unpleasant thing happening. Not the seat of fear itself, but of a prediction.[18] We know something of the spikes in an animal's amygdala, and of how they change during the learning of such a prediction.[19] But we know nothing of the spikes that give the subjective experience of "fear." Nor happiness. Nor ambivalence. Indeed, Lisa Feldman Barrett could write an intriguing 411-page book about emotion and the brain without mentioning spikes once.

The explanatory gap between spikes and subjective experience also manifests in the vestiges of Cartesian dualism, the idea that mind and brain are somehow separate. Especially in the ultimate subjective experience of consciousness. The ephemeral sensation of being aware of one's self, the inner monologue, the experience of things—of redness, crunchiness, stickiness, of taste, texture, and smell. Nothing is known of how spikes passed between neurons relate to consciousness. This lack of grounding in the actual mechanics of the brain sets consciousness research adrift.

Scientists studying consciousness have to get around the problem that we know nothing of what signals the neurons are sending. Some look to fMRI, to at least catch a glimpse of which giant groups of neurons may be increasing or decreasing their activity during subjective experience.[20] Some look to the wiring between the regions of the human brain and argue for its unparalleled complexity.[21] Some have taken the giant leap of skipping over spikes entirely, not to mention the complex computations a single neuron can carry out both through its dendrites and through its mind-bogglingly complex chemical signaling pathways, and diving straight down to the quantum level[22]—which seems odd, given that perhaps we ought to first test a theory of consciousness at the level that we know information is passed in the brain.

Some philosophers studying consciousness fall headlong into this explanatory gap. For some defend dualism, reasoning that because we can't find a physical explanation for consciousness, nor can we think of one, the mind thus has no physical essence.[23] There's a simple reason for why we can't find or think of one. These vestiges of dualism live on because we cannot yet link the actions of individual neurons to these subjective mental states. Not because there is no link, but because we don't have, and as yet cannot get, the necessary data. We've literally never tried to link the activity of your neurons, the legion of spikes, all two billion of them a second, to your conscious experience.

The future of spikes is to ground idle speculation in knowledge. More spikes is a given, and from those we will learn much more about all the phenomena we met on our journey through your brain, about the legion and the meaning of spike, about the dark neurons and the spontaneous spikes. The more spikes we can gather, the more we will learn by omission about the elements of the brain that are not controlled by spikes, about the things that we cannot explain no matter how many spikes we greedily snap up—like mood, and perhaps memories of the distant past. And we will almost certainly get different spikes, spikes from disorders we know nothing about, from human thought processes we have yet to touch on, from subjective experiences we have yet to record a single spike during—spikes that will enrich our understanding of what it means to be human. That's where we ought to journey next.

ACKNOWLEDGMENTS

As a kid I wrote a lot of chapter 1s, excitedly scribbling the first install-
ment of an epic tale to be told, only for the momentum to ebb away, the
thrill fizzling out after the initial rush. I never made it to chapter 2. It
turned out that the best way to make it all the way to chapter 11 was to
combine two simple ingredients: a plan of what you're going to write
and, just as importantly, the support of a wonderful cast of people. It is
them I want to thank here.

My agent Jeff Shreve, of The Science Factory, for provoking this entire project, playing a crucial role in shaping the initial ideas for this
book, and his wise counsel throughout. The wonderful team at Princeton University Press, especially my editor Hallie Stebbins for bringing
this project to Princeton and her guiding hand, and Dawn Hall for her
diligent copyediting and determination to wean me off my addiction to
semicolons; sadly, not entirely with success.

I am indebted to my scientific friends and colleagues around the
world who gave their time and expertise to help improve this book.
Riccardo Storchi and Tim Vogels gave me nuggets of information that
played key roles. Tom Baden, Tiago Branco, Matteo Carandini, Mark
Churchland, and Rasmus Petersen all gave me the benefit of their deep
expertise by reading sections of the book, catching my errors and pointing me to profitable new avenues to explore. Ashley Juavinett's deep
reading of the first four chapters improved them considerably. And I
commend Pat Scannell and my reviewers at Princeton University Press
(Matthias Hennig and his anonymous counterpart) for taking on the
whole thing, and helping me push this book on—to the extent of writing a whole new chapter at the end. Thank you all. (And, Pat, I'm sorry
I couldn't find room in a book about neurons for your frankly terrifying

anecdote that there are more armed hunters out on a weekend in Wisconsin than there are combatants in the combined armed forces of France and Germany. So here it is.)

My kids, Abbi and Seth, while dimly aware that Daddy was writing a book about brains, rightly thought they had more interesting things to keep Daddy occupied with—including teddy hospitals, football in the garden, and bedtime stories—and so kept Daddy sane. And for her unstinting support throughout, my deepest thanks and love to my wife Nic.

NOTES

CHAPTER 1. INTRODUCTION

1. All neuron numbers are from Suzana Herculano-Houzel, *The Human Advantage*, MIT Press, 2016.

2. Peter Lennie, "The cost of cortical computation," *Current Biology* 13 (2003): 493–97; Simon B. Laughlin and Terry J. Sejnowski, "Communication in neuronal networks," *Science* 301 (2003): 1870–74.

3. I estimated 450 trillion words in all published English novels, given the total number of published English novels is about five million (Fredner), and the average word count per novel is ninety thousand. Estimates of about 100,000 new English novels per year were used to estimate the 380 million years needed to reach parity between words and spikes.

Erik Fredner, "How many novels have been published in English? (An attempt)," March 14, 2017, https://litlab.stanford.edu/how-many-novels-have-been-published-in-english-an-attempt/.

4. One of the first reports of recording spikes was Edgar D. Adrian, "The impulses produced by sensory nerve endings: Part I," *Journal of Physiology* 61 (1926): 49–72.

5. This is the neuroscience equivalent of Moore's law: the technology for recording spikes from many neurons at the same time is exponentially increasing the number of neurons it can record over time, doubling about every 6.3 years. We might dub this "Stevenson's Law": Ian Stevenson and Konrad Kording, "How advances in neural recording affect data analysis," *Nature Neuroscience* 14 (2011): 139–42; and Ian Stevenson's website: https://stevenson.lab.uconn.edu/scaling/.

6. Turning neurons on or off with light is achieved using optogenetics, first used in mammals in 2005. Some bacteria have ion channels in their skin that open when light falls on them. By inserting the genes for those ion channels into a neuron, we can make that neuron form the same ion channels in its skin. Then when we shine light on the neuron, the channel opens, and ions rush in or out (depending on which type of ion channel is expressed by the genes), either exciting that neuron or inhibiting it. This can be done in thousands of neurons at once and in specific types of neurons.

For more, see: Gero Miesenböck, "The optogenetic catechism," *Science* 326 (2009): 395–99; Karl Deisseroth, "Optogenetics: 10 years of microbial opsins in neuroscience," *Nature Neuroscience* 18 (2015): 1213–25.

7. Examples of recording spikes in humans from the deep brain stimulation electrode: René Reese, Arthur Leblois, Frank Steigerwald, Monika Pötter-Nergera, Jan Herzoga, H. Maximilian

Mehdorn, Günther Deuschl, Wassilios G. Meissnerd, and Jens Volkmann, "Subthalamic deep brain stimulation increases pallidal firing rate and regularity," *Experimental Neurology* 229 (2011): 517–21; Arun Singh, Klaus Mewes, Robert E. Gross, Mahlon R. DeLong, José A Obeso, and Stella M. Papa, "Human striatal recordings reveal abnormal discharge of projection neurons in Parkinson's disease," *Proceedings of the National Academy of Sciences USA* 113 (2016): 9629–34.

8. For an example of recording spikes from electrodes implanted to find the focus of epileptic seizures in human patients, see Matias J. Ison, Rodrigo Quian Quiroga, and Itzhak Fried, "Rapid encoding of new memories by individual neurons in the human brain," *Neuron* 87 (2015): 220–30.

9. R. Jenkins, A. J. Dowsett, and A. M. Burton, "How many faces do people know?" *Proceedings of the Royal Society B: Biological Sciences* 285 (2018): https://royalsocietypublishing.org/doi/10.1098/rspb.2018.1319.

10. Nancy Kanwisher, Josh McDermott, and Mavin M. Chun, "The fusiform face area: A module in human extrastriate cortex specialized for face perception," *Journal of Neuroscience* 17 (1997): 4302–11.

11. For a lovely account of Doris Tsao's work, and its context within other work on cracking the face code, see Alison Abbott, "The face detective," *Nature* 564 (2018): 176–79.

12. Doris Y. Tsao, Winrich A. Freiwald, Roger B. H. Tootell, and Margaret S. Livingstone, "A cortical region consisting entirely of face-selective cells," *Science* 311 (2006): 670–74.

13. Sebastian Moeller, Winrich A. Freiwald, and Doris Y. Tsao, "Patches with links: A unified system for processing faces in the macaque temporal lobe," *Science* 320 (2008): 1355–59.

14. Le Chang and Doris Y. Tsao, "The code for facial identity in the primate brain," *Cell* 169 (2017): 1013–28.

CHAPTER 2. ALL OR NOTHING

1. Warren McCulloch is a fascinating early pioneer of systems neuroscience and computational neuroscience. A great biographical sketch, from someone who knew McCulloch in his later years, is Arbib. A fuller biography of McCulloch, setting his work in the context of the time, can be found in Abraham.

Michael A. Arbib, "Warren McCulloch's search for the logic of the nervous system," *Perspectives in Biology and Medicine* 43 (2000): 193–16.

Tara H. Abraham, *Rebel Genius: Warren S. McCulloch's Transdisciplinary Life in Science*, MIT Press, 2016.

2. If they were lucky. Oscilloscopes didn't come into common usage until the 1930s. So the very earliest recordings of spikes, like Edgar Adrian's, used hand-built contraptions to convert these minuscule flickers of voltage into pen movements on a rolling chart.

3. This account of the early days of spike recordings up to Hodgkin and Huxley's work in the 1950s draws on Alan J. McComas, *Galvani's Spark: The Story of the Nerve Impulse*, Oxford University Press, 2011.

4. The use of "tipping point" here is not accidental. Readers familiar with the basics of neuroscience may be wondering why I don't say "threshold," like every textbook account of how a neuron creates a spike: "and, lo, when the voltage reaches a threshold, a spike is born." That's because a threshold, in the sense of an exact amount of voltage at which a spike will be created,

does not exist. The voltage at which a neuron will make a spike depends on what else has happened to the neuron recently—most importantly, how long ago its last spike was. So there is always a voltage at which a spike will be made—a tipping point. But not always the same voltage—no threshold. For a book-length explanation of why neurons do not have thresholds, see the dense but awe-inspiring Eugene Izhikevich, *Dynamical Systems in Neuroscience: The Geometry of Excitability and Bursting*, MIT Press, 2005. Or to find out how variable the tipping point can be, see, for example, Johnathan Platkiewicz and Romain Brette, "A threshold equation for action potential initiation," *PLoS Computational Biology* 6 (2010): e1000850.

5. Amanda Gefter, "The man who tried to redeem the world with logic," *Nautilus*, February 5, 2015, http://nautil.us/issue/21/information/the-man-who-tried-to-redeem-the-world-with-logic; Neil R. Smalheiser, "Walter Pitts," *Perspectives in Biology and Medicine* 43 (2000): 217–26.

6. John von Neumann, "First draft of a report on the EDVAC," 1945 Technical Report. Typeset and edited by Michael D Godfrey; version dated January 10, 2011. Downloaded from https://sites.google.com/site/michaeldgodfrey/vonneumann/vnedvac.pdf?attredirects=0&d=1.

7. Indeed, the circuits formed between the three layers of the retina are so ridiculously complex, it gives one pause for thought about our ability to understand the vastly larger brain. For those who like the technical details, here's a crash course on the retina. The cones on the first layer release glutamate onto the bipolar cells in the second layer, and also onto the "horizontal" cells there. The job of the bipolar cells is to compress down the information from the cones and pass it on to the third layer. The job of the horizontal cells is to inhibit transmission from the cones to the bipolar cells; at sites farther away from the cones they get input from themselves. This creates competition: it means the response of the bipolar cells farther away is suppressed, so that the response of the bipolar cells near the active horizontal cell(s) stands out.

The bipolar cells' voltage changes in proportion to the size of the pause in glutamate they receive from the type of cone(s) they are coupled to. The ON type of bipolar cells increase their voltage in response to the pauses, signaling the detection of photons; OFF bipolar cells decrease their voltage in response to the pauses, signaling a decrease in darkness. In turn, all bipolar cells release glutamate in proportion to their voltage onto the neurons in the third layer—the retinal ganglion cells (the ones that transmit spikes to the brain) and the amacrine cells (that inhibit the ganglion cells and/or feedback inhibition to the bipolar cells).

In the mouse retina, there are at least nine types of bipolar cell, and at least forty types of amacrine cell. These types are defined by what they respond to. For a full technical account of the retina's circuitry, see Jonathan B. Demb and Joshua H. Singer, "Functional circuitry of the retina," *Annual Review of Vision Science* 1 (2015): 263–69. The amount we know about bipolar cells is absurd: see Thomas Euler, Silke Haverkamp, Timm Schubert, and Tom Baden, "Retinal bipolar cells: Elementary building blocks of vision," *Nature Reviews Neuroscience* 15 (2014): 507–19. For a detailed account of why the retina is wired this way, see chapter 11 of Peter Sterling and Simon B. Laughlin, *Principles of Neural Design*, MIT Press, 2015. A useful place to get started is the online textbook *Webvision*, by Helga Kolb and her colleagues: https://webvision.med.utah.edu/.

8. Amanda Gefter, "The man who tried to redeem the world with logic," *Nautilus*, February 5, 2015, http://nautil.us/issue/21/information/the-man-who-tried-to-redeem-the-world-with-logic.

9. Michael Brecht, Bruno Preilowski, and Michael M. Merzenich, "Functional architecture of the mystacial vibrissae," *Behavioural Brain Research* 84 (1997): 81–97.

10. Michael R. Bale, Dario Campagner, Andrew Erskine, and Rasmus S. Petersen, "Microsecond-scale timing precision in rodent trigeminal primary afferents," *Journal of Neuroscience* 35 (2015): 5935–40.

11. Magdalena N. Muchlinski, John R. Wible, Ian Corfe, Matthew Sullivan, and Robyn A. Grant, "Good vibrations: The evolution of whisking in small mammals," *Anatomical Record* 303 (2020): 89–99.

12. Ben Mitchinson, Chris J. Martin, Robyn A. Grant, and Tony J. Prescott, "Feedback control in active sensing: Rat exploratory whisking is modulated by environmental contact," *Proceedings of the Royal Society B: Biological Sciences* 27 (2007): 1035–41; Robyn A. Grant, Ben Mitchinson, Charles W. Fox, and Tony J. Prescott, "Active touch sensing in the rat: Anticipatory and regulatory control of whisker movements during surface exploration," *Journal of Neurophysiology* 101 (2009): 862–74.

13. Rune W. Berg and David Kleinfeld, "Rhythmic whisking by rat: Retraction as well as protraction of the vibrissae is under active muscular control," *Journal of Neurophysiology* 89 (2003): 104–17.

14. This is an "as the crow flies" estimate: the length of the human brain is around 150–160 millimeters. Of course, a spike could not travel in that straight line—the imaginary axon carrying that spike from the back to the front of the brain would first turn downward to enter the white matter underneath the cortex, then travel in a curved trajectory through the white matter before turning back up into the cortex at the front of the brain. Chapter 4 will elaborate.

15. Heather L. More, John R. Hutchinson, David F. Collins, Douglas J. Weber, Steven K. H. Aung, and J. Maxwell Donelan, "Scaling of sensorimotor control in terrestrial mammals," *Proceedings of the Royal Society B: Biological Sciences* 277 (2010): 3563–68.

16. Peter Sterling and Simon Laughlin, *Principles of Neural Design*, MIT Press, 2015, chapter 7.

17. The estimate of how many cortical cells lie end-to-end across the length of the cortex relies on data obtained in the rat (Romand et al.). For two reasons: (1) we have lots of accurate data for rats; (2) rat cortex is not folded, unlike the human cortex, so straight line distances are meaningful. That human cortex is folded means that there are likely far more neurons in the sequence from back to front than estimated here. The numbers: length of the cortex in adult rats is ~14 mm. The tufted pyramidal neurons of layer five are the largest neurons, and their bodies are up to 20 micrometers across. So we can fit 700 of these neuron bodies adjacent to each other in the distance from the front to the back of the cortex.

Sandrine Romand, Yun Wang, Maria Toledo-Rodriguez, and Henry Markram, "Morphological development of thick-tufted layer V pyramidal cells in the rat somatosensory cortex," *Frontiers in Neuroanatomy* 5 (2011): 5.

18. Robyn A. Grant, Vicki Breakell, and Tony J. Prescott, "Whisker touch sensing guides locomotion in small, quadrupedal mammals," *Proceedings of the Royal Society B: Biological Sciences* 285 (2018): 20180592.

19. A neuron with a two-meter-long axon is sending a signal farther, relative to its own size, than the distance from the Earth to the Sun. Distance from the Earth to the Sun averages

149,600,000 km, and Earth's diameter at the equator is 12,756 km. So the distance from the Earth to the Sun is about 11,723 Earths.

Now take a neuron sending a two-meter-long axon. Say its body is about 20 micrometers in diameter (which is generous): then its axon is 100,000 times longer than the neuron's body is wide. (If you're the kind of pedantic person who wants to count not just the neuron's body but also its dendrites within its diameter, then if that was, say, 200 micrometers, the axon is 10,000 times longer than the neuron is wide.)

20. There is a downside to the eye translating all this voltage and chemistry into spikes: losing information. The cortex receives a mere fraction of the information about the visible world available from the photons landing on the cones of the retina. We can think of the messages passed from the cones to the output neurons of the retina, the ganglion cells, as analogue: they are continuous changes in voltage that in turn regulate the flow of transmitters. But in translating these continuous changes into spikes, the ganglion cells are forced to discard much of the detail of those messages. They quantize the messages, turning a continuous signal into the occasional binary event—a spike. Each spike thus stands for a range of possible values, rather than one. Inevitably then, information is lost in the process. Which means that in receiving these spikes, the visual cortex is missing a lot of potential information about the world outside.

21. J. Y. Lettvin, H. R. Maturana, W. S. McCulloch, and W. H. Pitts, "What the frog's eye tells the frog's brain," *Proceedings of the Institute of Radio Engineers (IRE)* 47 (1959): 1940–51.

You may note that Walter Pitts's name appears on the author list of this paper, though as Lettvin recorded the neural activity, and Maturana was responsible for making sense of the anatomy, the role of Pitts in his own downfall is unclear.

22. Tom Baden, Philipp Berens, Katrin Franke, Miroslav Román Rosón, Matthias Bethge, and Thomas Euler, "The functional diversity of retinal ganglion cells in the mouse," *Nature* 529 (2016): 345–50.

23. Peter Sterling and Simon Laughlin, *Principles of Neural Design*, MIT Press, 2015, chapter 11. More precisely: the direction-selective ON cells are used to compute the slippage of the scene on your retina as the head moves.

24. Tom Baden, Thomas Euler, and Philipp Berens, "Understanding the retinal basis of vision across species," *Nature Reviews Neuroscience* 21 (2020): 5–20.

25. Bruce A. Rheaume, Amyeo Jereen, Mohan Bolisetty, Muhammad S. Sajid, Yue Yang, Kathleen Renna, Lili Sun, Paul Robson, and Ephraim F. Trakhtenberg, "Single cell transcriptome profiling of retinal ganglion cells identifies cellular subtypes," *Nature Communications* 9 (2018): 2759; Yi-Rong Peng, Karthik Shekhar, Wenjun Yan, Dustin Herrmann, Anna Sappington, Gregory S. Bryman, Tavévan Zyl, Michael Tri H. Do, Aviv Regev, and Joshua R. Sanes, "Molecular classification and comparative taxonomics of foveal and peripheral cells in primate retina," *Cell* 176 (2019): 1222–37.

CHAPTER 3. LEGION

1. Some of you are thinking: hey, Humphries, whatever happened to the lateral geniculate nucleus? Spikes do not go directly from the retina to the first visual region of the cortex. The axons of retinal ganglion cells arrive in the lateral geniculate nucleus (LGN), part of the thalamus. One set of neurons there relays the retina's signal to the visual bits of cortex; a separate set

of LGN neurons relay the retina's signal to numerous structures below the cortex, such as the superior colliculus, for rapid responses to changes in your field of view (like ducking an incoming football). Given the thirty channels coming out of the retina, there is, as you might imagine, an equally complex set of channels coming out of the LGN, which are just being revealed. See Miroslav Román Rosón, Yannik Bauer, Ann H. Kotkat, Philipp Berens, Thomas Euler, and Laura Busse, "Mouse dLGN receives functional input from a diverse population of retinal ganglion cells with limited convergence," *Neuron* 102 (2019): 462–76.

2. Nuno Macarico da Costa and Kevin A. C. Martin, "How thalamus connects to spiny stellate cells in the cat's visual cortex," *Journal of Neuroscience* 31 (2011): 2925–37.

3. Neurons come in a staggering array of shapes. Bountiful examples of the shapes formed by these cortical neurons, and many other neuron types, are freely available at NeuroMorpho .org.

4. Well, almost never is one spike arriving enough to cause a spike in its target neuron. But as we know, nature likes to make a mockery of our attempts to create ironclad laws. A single axon can make multiple contacts with the tree of a single target neuron. Each of these gaps thus gets the same spike, and so a single spike can make multiple increases (or decreases) in voltage at the same time. The extreme case is the "detonator synapse" made by a neuron from the dentate gyrus onto a neuron in area CA3 (both are bits of the hippocampus). The single gyrus neuron makes many, strong connections with the CA3 neuron. Under the right circumstances (namely, these connections recently had a lot of spikes passing through them), these connections can be made so strong that a single spike from the gryus neuron causes a spike in the CA3 neuron. For the crucial experiments, see Nicholas P. Vyleta, Carolina Borges-Merjane, and Peter Jonas, "Plasticity-dependent, full detonation at hippocampal mossy fiber-CA3 pyramidal neuron synapses," *eLife* 5 (2016): e17977. For a nuanced, quantitative discussion of why this must be a rare event, see Nathaniel N. Urban, Darrell A. Henze, and German Barrionuevo, "Revisiting the role of the hippocampal mossy fiber synapse," *Hippocampus* 11 (2001): 408–17.

5. For any serious student of the cortex, Braitenberg and Schuz's monograph on their work to unpick the statistics of the cortex is essential reading. Valentin Braitenberg and Almut Schuz, *Cortex: Statistics and Geometry of Neuronal Connectivity*, 2nd ed., Springer, 1998.

6. Michael London, Arnd Roth, Lisa Beeren, Michael Häusser, and Peter E. Latham, "Sensitivity to perturbations in vivo implies high noise and suggests rate coding in cortex," *Nature* 466 (2010): 123–27.

7. Michelle Rudolph and Alain Destexhe, "Tuning neocortical pyramidal neurons between integrators and coincidence detectors," *Journal of Computational Neuroscience* 14 (2003): 239–51.

8. Mark D. Humphries, "The Goldilocks zone in neural circuits," *eLife* 5 (2016): e22735.

9. William R. Softky and Christof Koch, "Cortical cells should fire regularly, but do not," *Neural Computation* 4 (1992): 643–46.

10. By "perfectly random" I mean that the trains of spikes followed a Poisson process, as though the gaps between spikes were randomly drawn from an exponential distribution, and each draw is completely independent. To be a Poisson process, the mean and standard deviation of the gaps between spikes should be the same—and Softky and Koch showed they were (roughly).

11. This chapter outlines the currently favored "balanced-input" theory of cortex. But other proposals for how irregular spikes may arise contributed to this theory too. Softky and Koch's own solution [1] was that our neuron models were too simple, and this irregular output came from the trees of the target neuron filtering out small, isolated inputs, and only sending a big voltage blip to the body when enough inputs turned up at once, which would happen randomly. Others showed that tweaking simple features of our neuron models could also make the gaps between spikes more irregular: like increasing how much the voltage plunged downward after a spike [2], or making the delay in starting to spike depend on how quickly the voltage reached the neuron's tipping point [3].

[1] William R. Softky and Christof Koch, "The highly irregular firing of cortical cells is inconsistent with temporal integration of random EPSPs," *Journal of Neuroscience* 13 (1993): 334–50.

[2] Todd T. Troyer and Kenneth D. Miller, "Physiological gain leads to high ISI variability in a simple model of a cortical regular spiking cell," *Neural Computation* 9 (1997): 971–83.

[3] Boris Gutkin and G. Bard Ermentrout, "Dynamics of membrane excitability determine interspike interval variability: A link between spike generation mechanisms and cortical spike train statistics," *Neural Computation* 10 (1998): 1047–65.

12. Shadlen and Newsome's initial ideas were published in 1994, eighteen months or so after Softky and Koch's first paper, with the full models following in 1998.

Michael N. Shadlen and William T. Newsome, "Noise, neural codes and cortical organization," *Current Opinion in Neurobiology* 4 (1994): 569–79; Michael N. Shadlen and William T. Newsome, "The variable discharge of cortical neurons: Implications for connectivity, computation, and information coding," *Journal of Neuroscience* 18 (1998): 3870–96.

13. Alain Destexhe, Michelle Rudolph, and Denis Paré, "The high-conductance state of neocortical neurons in vivo," *Nature Reviews Neuroscience* 4 (2003): 739–51.

14. Misha V. Tsodyks and Terry Sejnowski, "Rapid state switching in balanced cortical network models," *Network* 6 (1995): 111–24.

15. Carl van Vreeswijk and Haim Sompolinsky, "Chaos in neuronal networks with balanced excitatory and inhibitory activity," *Science* 274 (1996): 1724–26; Carl van Vreeswijk and Haim Sompolinsky, "Chaotic balanced state in a model of cortical circuits," *Neural Computation* 10 (1998): 1321–71.

16. Christopher I. Moore and Sacha B. Nelson, "Spatio-temporal subthreshold receptive fields in the vibrissa representation of rat primary somatosensory cortex," *Journal of Neurophysiology* 80 (1998): 2882–92.

17. Yousheng Shu, Andrea Hasenstaub, and David A. McCormick, "Turning on and off recurrent balanced cortical activity," *Nature* 423 (2003): 288–93; Bilal Haider, Alvaro Duque, Andrea R. Hasenstaub, and David A, McCormick, "Neocortical network activity in vivo is generated through a dynamic balance of excitation and inhibition," *Journal of Neuroscience* 26 (2006): 4535–45.

18. Michael Wehr and Anthony M. Zador, "Balanced inhibition underlies tuning and sharpens spike timing in auditory cortex," *Nature* 426 (2003): 442–46.

19. Michael Okun and Ilan Lampl, "Instantaneous correlation of excitation and inhibition during ongoing and sensory-evoked activities," *Nature Neuroscience* 11 (2008): 535–37.

20. Tal Kenet, Amos Arieli, Misha Tsodyks, and Amiram Grinvald, "Are single neurons soloists or are they obedient members of a huge orchestra?" in *23 Problems in Systems Neuroscience*, ed. J. L. van Hemmen and T. J. Sejnowski, Oxford University Press, 2006, 160–81.

21. Michael Okun, Nicholas A. Steinmetz, Lee Cossell, M. Florencia Iacaruso, Ho Ko, Péter Barthó, Tirin Moore, Sonja B. Hofer, Thomas D. Mrsic-Flogel, Matteo Carandini, and Kenneth D. Harris, "Diverse coupling of neurons to populations in sensory cortex," *Nature* 521 (2015): 511–15.

22. Matteo Carandini, Jonathan B. Demb, Valerio Mante, David J. Tolhurst, Yang Dan, Bruno A. Olshausen, Jack L. Gallant, and Nicole C. Rust, "Do we know what the early visual system does?" *Journal of Neuroscience* 25 (2005): 10577–97.

23. Cyrille Rossant, Sara Leijon, Anna K. Magnusson, and Romain Brette, "Sensitivity of noisy neurons .to coincident inputs," *Journal of Neuroscience* 31 (2011): 17193–206.

24. Charles F. Stevens and Anthony M. Zador, "Input synchrony and the irregular firing of cortical neurons," *Nature Neuroscience* 1 (1998): 210–17.

25. Emilio Salinas and Terrence J. Sejnowski, "Impact of correlated synaptic input on output firing rate and variability in simple neuronal models," *Journal of Neuroscience* 20 (2000): 6193–209.

26. For examples of the tight delay between excitation and its balancing inhibition, see Michael Wehr and Anthony M. Zador, "Balanced inhibition underlies tuning and sharpens spike timing in auditory cortex," *Nature* 426 (2003): 442–46; Michael Okun and Ilan Lampl, "Instantaneous correlation of excitation and inhibition during ongoing and sensory-evoked activities," *Nature Neuroscience* 11 (2008): 535–37.

27. For an accessible review of how dendrites could compute, start with Michael London and Michael Häusser, "Dendritic computation," *Annual Review of Neuroscience* 28 (2005): 503–32.

28. Lucy M. Palmer, Adam S. Shai, James E. Reeve, Harry L. Anderson, Ole Paulsen, and Matthew E. Larkum, "NMDA spikes enhance action potential generation during sensory input," *Nature Neuroscience* 17 (2014): 383–90.

29. Monika Jadi, Alon Polsky, Jackie Schiller, and Bartlett W. Mel, "Location-dependent effects of inhibition on local spiking in pyramidal neuron dendrites," *PLoS Computational Biology* 8 (2012): e1002550.

30. The equivalence between a pyramidal neuron and a two-layer neural network was established in a pair of papers published back to back: Panayiota Poirazi, Terrence Brannon, and Bartlett W. Mel, "Arithmetic of subthreshold synaptic summation in a model CA1 pyramidal cell," *Neuron* 37 (2003): 977–87; Panayiota Poirazi, Terrence Brannon, and Bartlett W. Mel, "Pyramidal neuron as two-layer neural network," *Neuron* 37 (2003): 989–99.

31. Romain D. Cazé, Mark Humphries, and Boris Gutkin, "Passive dendrites enable single neurons to compute linearly non-separable functions," *PLoS Computational Biology* 9 (2013): e1002867.

32. Mark Humphries, "Your cortex contains 17 billion computers," *The Spike*, February 12, 2018, https://medium.com/the-spike/your-cortex-contains-17-billion-computers -9034e42d34f2.

CHAPTER 4. SPLIT PERSONALITY

1. Ho Ko, Sonja B. Hofer, Bruno Pichler, Katherine A. Buchanan, P. Jesper Sjöström, and Thomas D. Mrsic-Flogel, "Functional specificity of local synaptic connections in neocortical networks," *Nature* 473 (2011): 87–91.

2. Lee Cossell, Maria Florencia Iacaruso, Dylan R. Muir, Rachael Houlton, Elie N. Sader, Ho Ko, Sonja B. Hofer, and Thomas D. Mrsic-Flogel, "Functional organization of excitatory synaptic strength in primary visual cortex," *Nature* 518 (2015): 399–403.

3. The Mrsic-Flogel lab also showed that the selective connections between similarly tuned neurons in the first bit of visual cortex (V1) seem to arise during development (Ko et al.). Early in development, connections between neurons are made at random. But the tuning-selective wiring appears by early adulthood (in mice), possibly by following a rule thought fundamental to brain function: neurons that fire together wire together. Here in V1, simple cells getting the same information in the same location from the retina will respond similarly to the world and so reinforce the wiring between them; by contrast, simple cells getting different information do not fire together, so the wiring between them atrophies. Result: simple cells reinforce each other.

By the way, this has a cool implication: that if you were to create unnatural statistics of the visual world, like making all edges horizontal, and allow a brain to develop in that world, then this would change which simple cells respond together, and so then should change wiring between them in V1.

Ho Ko, Lee Cossell, Chiara Baragli, Jan Antolik, Claudia Clopath, Sonja B. Hofer, and Thomas D. Mrsic-Flogel, "The emergence of functional microcircuits in visual cortex," *Nature* 496 (2013): 96–100.

4. For models of how complex cells in V1 are made by combining the outputs of simple cells, see Matteo Carandini, Jonathan B. Demb, Valerio Mante, David J. Tolhurst, Yang Dan, Bruno A. Olshausen, Jack L. Gallant, and Nicole C. Rust, "Do we know what the early visual system does?" *Journal of Neuroscience* 25 (2005): 10577–97; Nicole C. Rust, Odelia Schwartz, J. Anthony Movshon, and Eero P. Simoncelli, "Spatiotemporal elements of macaque V1 receptive fields," *Neuron* 46 (2005): 945–56.

5. Ferenc Mechler and Dario L. Ringach, "On the classification of simple and complex cells," *Vision Research* 42 (2002): 1017–33; Nicole C. Rust, Odelia Schwartz, J. Anthony Movshon, and Eero P. Simoncelli, "Spatiotemporal elements of macaque V1 receptive fields," *Neuron* 46 (2005): 945–56.

6. M. Florencia Iacaruso, Ioana T. Gasler, and Sonja B. Hofer, "Synaptic organization of visual space in primary visual cortex," *Nature* 547 (2017): 449–52.

7. Selmaan N. Chettih and Christopher D. Harvey, "Single-neuron perturbations reveal feature-specific competition in V1," *Nature* 567 (2019): 334–40.

8. Bilal Haider, Michael Häusser, and Matteo Carandini, "Inhibition dominates sensory responses in the awake cortex," *Nature* 493 (2013): 97–100.

9. For a review of the layers of cortex and the neuron types within them, see Kenneth D. Harris and Gordon M. G. Shepherd, "The neocortical circuit: Themes and variations," *Nature Neuroscience* 18 (2015): 170–81.

10. The idea of two separate streams of visual processing in the cortex is decades old. These map onto the so-called dorsal and ventral streams that are defined anatomically, what I'm calling respectively "Highway Do" and "Highway What." For overviews, see:

Melvyn A. Goodale and A. David Milner, "Separate visual pathways for perception and action," *Trends in Neurosciences* 15 (1992): 20–25.

Leslie G. Ungerleider and James V. Haxby, " 'What' and 'where' in the human brain," *Current Opinion in Neurobiology* 4 (1994): 157–65.

Melvyn A. Goodale, "How (and why) the visual control of action differs from visual perception," *Proceedings: Biological Sciences* 281 (2014): 20140337.

11. Geoffrey M. Boynton and Jay Hegdé, "Visual cortex: The continuing puzzle of area V2," *Current Biology* 14 (2004): R523–R524.

12. Edward H. Adelson, "On seeing stuff: The perception of materials by humans and machines," *Proceedings of the SPIE* 4299 (2001): 1–12.

13. Anthony J. Movshon and Eero P. Simoncelli, "Representation of naturalistic image structure in the primate visual cortex," *Cold Spring Harbor Symposia on Quantitative Biology* 7 (2014): 115–22.

14. S. Zeki, "Colour coding in the cerebral cortex: The reaction of cells in monkey visual cortex to wavelengths and colours," *Neuroscience* 9 (1983): 741–65.

15. Gregory Horwitz and Charles Hass, "Nonlinear analysis of macaque V1 color tuning reveals cardinal directions for cortical color processing," *Nature Neuroscience* 15 (2012): 913–19.

16. Vincent Walsh, "How does the cortex construct color?" *Proceedings of the National Academy of Sciences USA* 96 (1999): 13594–96.

17. For a review of the computational problem of object recognition, see James J. DiCarlo, Davide Zoccolan, and Nicole C. Rust, "How does the brain solve visual object recognition?" *Neuron* 73 (2012): 415–34.

18. J. Anthony Movshon and William T. Newsome, "Visual response properties of striate cortical neurons projecting to area MT in macaque monkeys," *Journal of Neuroscience* 16 (1996): 7733–41.

19. For models of how area MT neurons might compute global motion, see Eero P. Simoncelli and David J. Heeger, "A model of neuronal responses in visual area MT," *Vision Research* 38 (1998): 743–61; Nicole C. Rust, Valerio Mante, Eero P. Simoncelli, and J. Anthony Movshon, "How MT cells analyze the motion of visual patterns," *Nature Neuroscience* 9 (2006): 1421–31.

20. For an early example of deep networks modeling the visual system, see Honglak Lee, Chaitanya Ekanadham, and Andrew Y. Ng, "Sparse deep belief net model for visual area V2," in *Advances in Neural Information Processing Systems*, vol. 20, ed. J. C. Platt, D. Koller, Y. Singer, and S. T. Roweis, 2008, 873–80.

Note there is nothing special about "new" AI methods such as deep neural networks. This game of train-and-compare has been going on for the visual system since the 1980s. See, e.g., S. R. Lehky and T. J. Sejnowski, "Network model of shape-from-shading: Neural function arises from both receptive and projective fields," *Nature* 333 (1998): 452–54.

21. For a review of deep neural network modeling of vision, see Grace Lindsay, "Convolutional neural networks as a model of the visual system: Past, present, and future" (2020): https://arxiv.org/abs/2001.07092.

22. Jim DiCarlo's lab did this work on matching deep networks to the activity of temporal lobe neurons (Yamins et al.), and has recently pushed these deep neural networks to their limits (Bashivan, Kar, and DiCarlo). They first constructed synthetic images designed to control specific units or groups of units in the output layer of the network; they were constructed to either maximally increase the unit or group's activity or simultaneously increase the activity of one group and decrease the activity of another. They then showed these same synthetic images to monkeys, and the majority of monkeys' temporal lobe neurons responded in the exact same way as the deep network's—the ones meant to increase their activity mostly did so; the ones meant to decrease their activity mostly did so.

Daniel L. K. Yamins, Ha Hong, Charles F. Cadieu, Ethan A. Solomon, Darren Seibert, and James J. DiCarlo, "Performance-optimized hierarchical models predict neural responses in higher visual cortex," *Proceedings of the National Academy USA* 111 (2014): 8619–24.

Pouya Bashivan, Kohitij Kar, and James J. DiCarlo, "Neural population control via deep image synthesis," *Science* 364 (2019): eaav9436.

23. The classic paper on the hierarchy of the visual cortices is Daniel J. Felleman and David C. Van Essen, "Distributed hierarchical processing in the primate cerebral cortex," *Cerebral Cortex* 1 (1991): 1–47. See also David C. Van Essen, Charles H. Anderson, and Daniel J. Felleman, "Information processing in the primate visual system: An integrated systems perspective," *Science* 255 (1992): 419–23; Malcolm P. Young, "Objective analysis of the topological organization of the primate cortical visual system," *Nature* 358 (1992): 152–55.

For critiques of a purely hierarchical account, see, for example: Jay Hegdé and Daniel J. Felleman, "Reappraising the functional implications of the primate visual anatomical hierarchy," *Neuroscientist* 13 (2007): 416–21; S. Zeki, "The rough seas of cortical cartography," *Trends in Neurosciences* 41 (2018): 242–44.

24. Ricardo Gattass, Sheila Nascimento-Silva, Juliana G. M. Soares, Bruss Lima, Ana Karla Jansen, Antonia Cinira M. Diogo, Mariana F. Farias, Marco Marcondes, Eliã P. Botelho, Otávio S. Mariani, João Azzi, and Mario Fiorani, "Cortical visual areas in monkeys: Location, topography, connections, columns, plasticity, and cortical dynamics," *Philosophical Transactions of the Royal Society of London B: Biological Sciences* 360 (2005): 709–31.

25. The crucial nature of this feedback was emphasized in Felleman and Van Essen's classic paper on the visual hierarchy in cortex: Daniel J. Felleman and David C. Van Essen, "Distributed hierarchical processing in the primate cerebral cortex," *Cerebral Cortex* 1 (1991): 1–47.

26. Yunyun Han, Justus M. Kebschull, Robert A. Campbell, Devon Cowan, Fabia Imhof, Anthony M. Zador, and Thomas D. Mrsic-Flogel, "The logic of single-cell projections from visual cortex," *Nature* 556 (2018): 51–56.

27. For a recent example, see Simon Musall, Matthew T. Kaufman, Ashley L. Juavinett, Steven Gluf, and Anne K. Churchland, "Single-trial neural dynamics are dominated by richly varied movements," *Nature Neuroscience* 22 (2019): 1677–86.

28. Nuo Li, Kayvon Daie, Karel Svoboda, and Shaul Druckmann, "Robust neuronal dynamics in premotor cortex during motor planning," *Nature* 532 (2016): 459–64.

29. Kasper Winther Andersen and Hartwig Roman Siebner, "Mapping dexterity and handedness: Recent insights and future challenges," *Current Opinion in Behavioral Sciences* 20 (2018): 123–29.

30. S. Knecht, B. Dräger, M. Deppe, L. Bobe, H. Lohmann, A. Flöel, E-B. Ringelstein, and H. Henningsen, "Handedness and hemispheric language dominance in healthy humans," *Brain* 123 (2000): 2512–18.

31. Vyacheslav R. Karolis, Maurizio Corbetta, and Michel Thiebaut de Schotten, "The architecture of functional lateralisation and its relationship to callosal connectivity in the human brain," *Nature Communications* 10 (2019): 1417.

32. For a comprehensive review of the capabilities of split-brain patients, see Michael S. Gazzaniga, "Cerebral specialization and interhemispheric communication: Does the corpus callosum enable the human condition?" *Brain* 123 (2000): 1293–326.

CHAPTER 5. FAILURE

1. Anthony Zador, "Impact of synaptic unreliability on the information transmitted by spiking neurons," *Journal of Neurophysiology* 79 (1998): 1219–29; Amit Manwani and Christof Koch, "Detecting and estimating signals over noisy and unreliable synapses: Information-theoretic analysis," *Neural Computation* 13 (2001): 1–33.

2. Neal A. Hessler, Aneil M. Shirke, and Roberto Malinow, "The probability of transmitter release at a mammalian central synapse," *Nature* 366 (1993): 569–72.

3. Christina Allen and Charles F. Stevens, "An evaluation of causes for unreliability of synaptic transmission," *Proceedings of the National Academy of Sciences USA* 91 (1994): 10380–83.

4. For example, the synapse made by the neurogliaform interneuron on to the spiny projection neuron of the striatum was reported as having no failures (over the observed recording period). Osvaldo Ibáñez-Sandoval, Fatuel Tecuapetla, Bengi Unal, Fulva Shah, Tibor Koós, and James M. Tepper, "A novel functionally distinct subtype of striatal neuropeptide Y interneuron," *Journal of Neuroscience* 31 (2011): 16757–69.

5. Tiago Branco, Kevin Staras, Kevin J. Darcy, and Yukiko Goda, "Local dendritic activity sets release probability at hippocampal synapses," *Neuron* 59 (2008): 475–85.

6. The strength of a synapse, more formally. A spike arriving at a synapses releases N packets of transmitter molecules with probability p, and each packet of transmitter molecules has an effect q on the target neuron. These variables are the two parts of the "strength" of a synapse: $N \cdot q$ is the size of the voltage blip (well, technically, the conductance change), as more globs of transmitter means more activation of receptors, and p is how likely that blip is to happen. So the strength of a synapse (w) is proportional to the expected change it causes in the target neuron, i.e., $w \sim p \cdot (N \cdot q)$. So we can change a synapse's strength by changing either the size of the response (by changing N, changing q, or both), or by changing the probability of that response (p). See, e.g., chapters 4 and 13 in Christof Koch, *Biophysics of Computation*, MIT Press, 1999.

7. Tony M. Zador and Lynn E. Dobrunz, "Dynamic synapses in the cortex," *Neuron* 19 (1997): 1–4; Koch, *Biophysics of Computation*, chapter 13.

8. Charles F. Stevens and Yanyan Wang, "Changes in reliability of synaptic function as a mechanism for plasticity." *Nature* 371 (1994): 704–7.

9. William B. Levy and Robert A. Baxter, "Energy-efficient neuronal computation via quantal synaptic failures," *Journal of Neuroscience* 22 (2002): 4746–55.

10. The figure of 56 percent comes from Julia J. Harris, Renaud Jolivet, and David Attwell, "Synaptic energy use and supply," *Neuron* 75 (2012): 762–77.

11. Harris, Jolivet, and Attwell, "Synaptic energy use and supply."

12. Of course, there are other reasons why a neuron would make multiple contacts onto a target neuron. Prominent among these is because it wants to dominate control of that target neuron by contributing big jumps in the target neuron's voltage every time it fires a spike.

13. Tiago Branco, Kevin Staras, Kevin J. Darcy, and Yukiko Goda, "Local dendritic activity sets release probability at hippocampal synapses," *Neuron* 59 (2008): 475–85.

14. L. F. Abbott, J. A. Varela, K. Sen, and S. B. Nelson, "Synaptic depression and cortical gain control," *Science* 275 (1997): 220–24.

15. Charles F. Stevens and Tetsuhiro Tsujimoto, "Estimates for the pool size of releasable quanta at a single central synapse and for the time required to refill the pool," *Proceedings of the National Academy of Sciences USA* 92 (1995): 846–49.

16. Lynn E. Dobrunz and Charles F. Stevens, "Heterogeneity of release probability, facilitation, and depletion at central synapses," *Neuron* 18 (1997): 995–1008.

17. Wolfgang Maass, and Tony M. Zador, "Dynamic stochastic synapses as computational units," *Neural Computation* 11 (1999): 903–17.

18. The account I'm giving of how synaptic failure creates a filter for oscillations in the rate of spikes combines ideas from two papers. First, a classic paper on how short-term depression allows synapses to automatically control the gain of their synapses, which implies they filter high-frequency inputs: L. F. Abbott, J. A. Varela, K. Sen, and S. B. Nelson, "Synaptic depression and cortical gain control," *Science* 275 (1997): 220–24. And the working-out of that filter by Robert Rosenbaum, Jonathan Rubin, and Brent Doiron, "Short term synaptic depression imposes a frequency dependent filter on synaptic information transfer," *PLoS Computational Biology* 8 (2012): e1002557.

19. Mark D. Humphries, "The unreasonable effectiveness of deep brain stimulation," *The Spike*, January 30, 2017, https://medium.com/the-spike/the-unreasonable-effectiveness-of-deep-brain-stimulation-7d84a9849140.

20. Robert Rosenbaum, Andrew Zimnik, Fang Zheng, Robert S. Turner, Christian Alzheimer, Brent Doiron, and Jonathan E. Rubin, "Axonal and synaptic failure suppress the transfer of firing rate oscillations, synchrony and information during high frequency deep brain stimulation," *Neurobiology of Disease* 62 (2014): 86–99.

21. Dominic A. Evans, Vanessa Stempel, Ruben Vale, Sabine Ruehle, Yaara Lefler, and Tiago Branco, "A synaptic threshold mechanism for computing escape decisions," *Nature* 558 (2018): 590–94.

22. An intriguing aspect of these studies is that the escaping mice could run back to a shelter they had visited just once. So this seemingly instinctive behavior of "Flee!" actually uses some pretty advanced learning and planning. See Ruben Vale, Dominic A. Evans, and Tiago Branco, "Rapid spatial learning controls instinctive defensive behavior in mice," *Current Biology* 27 (2017): 1342–49.

23. Why neural networks overfit is exactly the same as why regression overfits data points: there are too many free parameters. In regression, you take a bunch of data points relating, say, household income to IMDb ratings for *Game of Thrones*, and fit a curve to them. That curve tells you the relationship between income and ratings. The more complex the curve, the better the fit to the existing data points, but more likely a worse fit to any new data points. And neural networks do the same thing: they fit highly complex curves to their inputs, to arrive at their outputs.

24. DropConnect was first reported in a conference paper (Wan et al.). When I say "widely used": this paper has 1,274 citations in six years at the time of writing (May 20, 2019; source: Google Scholar).

Li Wan, Matthew Zeiler, Sixin Zhang, Yann Le Cun, and Rob Fergus, "Regularization of neural networks using DropConnect," *Proceedings of the 30th International Conference on Machine Learning (ICML-13)* (2013): 1058–66.

I should also note that DropConnect builds on the idea of DropOut, which removed whole units at random from the network. The full paper on DropOut came out in 2014: Nitish Srivastava, Geoffrey Hinton, Alex Krizhevsky, Ilya Sutskever, and Ruslan Salakhutdinov, "DropOut: A simple way to prevent neural networks from overfitting," *Journal of Machine Learning Research* 15 (2014): 1929–58.

25. Some search algorithms will generate many initial solutions and refine them in parallel, and/or recombine them (as in evolutionary algorithms, such as genetic algorithms).

26. Why should we think the brain's search algorithm uses noise? Because neural networks that implement search algorithms also use noise—the most famous example is the "Boltzmann machine" (Ackley, Hinton, and Sejnowski). Way back in 1989, Burnod and Korn speculated that synaptic failure could be roughly akin to the noise in the Boltzmann machine. Their speculation rested on the probability of failure changing rapidly over time, to alter how a neuron responded to its inputs (high failure meaning it needed a lot of inputs). This seems unlikely at synapses within mammalian brains, but it was the first explicit link between failure and search.

David H. Ackley, Geoffrey E. Hinton, and Terrence J. Sejnowski, "A learning algorithm for Boltzmann machines," *Cognitive Science* 9 (1985): 147–69.

Y. Burnod and H. Korn, "Consequences of stochastic release of neurotransmitters for network computation in the central nervous system," *Proceedings of the National Academy of Sciences USA* 86 (1989): 352–56.

CHAPTER 6. THE DARK NEURON PROBLEM

1. Perhaps you're thinking—but what about synaptic failure? Wouldn't that bring down the total number of spikes arriving at a neuron? Indeed it would. But even if all the synapses on a single neuron had a failure rate of 50 percent, then that's still 25,000 input spikes a second, and a projected output rate of 250 spikes per second. And we were after 5 output spikes a second. To get from 50,000 input spikes per second to 5 output spikes per second would need a failure rate of 99 percent at every synapse.

2. To a first approximation, each spike sent by a neuron does indeed increase the amount of calcium floating freely in its body. And the chemical we use—the fluorescent indicator—binds calcium. The more it binds, the more it fluoresces, and so the brightness of fluorescence in the neuron's body is proportional to the number of spikes it sent. Roughly. In reality, there are problems. For one, the relationship between spikes and calcium is not linear; there is not the same increase in calcium for each additional spike. And another problem is that the changes in calcium are slow, much slower than a spike. So if multiple spikes are sent in a short time, the calcium signal is messy.

3. Jason N. D. Kerr, David Greenberg, and Fritjof Helmchen, "Imaging input and output of neocortical networks in vivo," *Proceedings of the National Academy of Science USA* 102 (2005): 14063–68.

4. Christopher D. Harvey, Philip Coen, and David W. Tank, "Choice-specific sequences in parietal cortex during a virtual-navigation decision task," *Nature* 484 (2012): 62–68.

5. Simon P. Peron, Jeremy Freeman, Vijay Iyer, Caiying Guo, and Karel Svoboda, "A cellular resolution map of barrel cortex activity during tactile behavior," *Neuron* 86 (2015): 783–99.

6. Tomáš Hromádka, Michael R. DeWeese, and Anthony M. Zador, "Sparse representation of sounds in the unanesthetized auditory cortex," *PLoS Biology* 6 (2008): e16.

7. Daniel H. O'Connor, Simon P. Peron, Daniel Huber, and Karel Svoboda, "Neural activity in barrel cortex underlying vibrissa-based object localization in mice," *Neuron* 67 (2010): 1048–61.

8. Alison L. Barth and James F. A. Poulet, "Experimental evidence for sparse firing in the neocortex," *Trends in Neurosciences* 35 (2012): 345–55.

9. For an early hint of the prevalence of dark neurons, see the discussion of David A. Robinson, "The electrical properties of metal microelectrodes," *Proceedings of the IEEE* 56 (1968): 1065–71.

The figure of about one hundred neurons in the recording radius of an electrode comes from D. A. Henze, Z. Borhegyi, J. Csicsvari, A. Mamiya, K. D. Harris, and G. Buzsáki, "Intracellular features predicted by extracellular recordings in the hippocampus in vivo," *Journal of Neurophysiology* 84 (2000): 390–400.

10. For a summary of the early evidence for dark neurons, see the review from 2006 that sparked my interest in the subject: Shy Shoham, Daniel H. O'Connor, and Ronen Segev, "How silent is the brain: Is there a 'dark matter' problem in neuroscience?" *Journal of Comparative Physiology A* 192 (2006): 777–84.

11. I note with some amusement that the dark neurons accidentally coincide with the old canard that we "only use 10 percent of our brains." That, of course, is total rubbish—as is crystal clear from fMRI studies showing blood flow everywhere in the brain, all the time. But in any given second, it does indeed appear that just 10 percent (roughly) of your cortical neurons are active.

12. Adrien Wohrer, Mark D. Humphries, and Christian Machens, "Population-wide distributions of neural activity during perceptual decision-making," *Progress in Neurobiology* 103 (2013): 156–93.

13. For an energy budget for the cortex, see Peter Lennie, "The cost of cortical computation," *Current Biology* 13 (2003): 493–97.

14. The classic paper on the use of energy by the gray matter of cortex is David Attwell and Simon B. Laughlin, "An energy budget for signaling in the grey matter of the brain," *Journal of Cerebral Blood Flow and Metabolism* 21 (2001): 1133–45. For more recent updates, see Biswa Sengupta, Martin Stemmler, Simon B. Laughlin, and Jeremy E. Niven, "Action potential energy efficiency varies among neuron types in vertebrates and invertebrates," *PLoS Computational Biology* 6 (2010): e1000840; Julia J. Harris, Renaud Jolivet, and David Attwell, "Synaptic energy use and supply," *Neuron* 75 (2012): 762–77.

15. Bruno A. Olshausen and David J. Field, "What is the other 85% of V1 doing?" in *23 Problems in Systems Neuroscience*, ed. J. L. van Hemmen and T. J. Sejnowski, Oxford University Press, 2006.

16. L. F. Abbott, J. A. Varela, K. Sen, and S. B. Nelson, "Synaptic depression and cortical gain control," *Science* 275 (1997): 220–24.

17. D. Huber, D. A. Gutnisky, S. Peron, D. H. O'Connor, J. S. Wiegert, L. Tian, T. G. Oertner, L. L. Looger, and K. Svoboda, "Multiple dynamic representations in the motor cortex during sensorimotor learning," *Nature* 484 (2012): 473–78.

18. Jose M. Carmena, Mikhail A. Lebedev, Craig S. Henriquez, and Miguel A. L. Nicolelis, "Stable ensemble performance with single-neuron variability during reaching movements in primates," *Journal of Neuroscience* 25 (2005): 10712–16.

19. Evan S. Hill, Sunil K. Vasireddi, Angela M. Bruno, Jean Wang, and William N. Frost, "Variable neuronal participation in stereotypic motor programs," *PLoS One* 7 (2012): e40579; Evan S. Hill, Sunil K. Vasireddi, Jean Wang, Angela M. Bruno, and William N. Frost, "Memory formation in tritonia via recruitment of variably committed neurons," *Current Biology* 25 (2015): 2879–88; Angela M. Bruno, William N. Frost, and Mark D. Humphries, "A spiral attractor network drives rhythmic locomotion," *eLife* 6 (2017): e27342.

20. Silvia Maggi, Adrien Peyrache, and Mark D. Humphries, "An ensemble code in medial prefrontal cortex links prior events to outcomes during learning," *Nature Communications* 9 (2018): 2204.

21. Wohrer, Humphries, and Machens, "Population-wide distributions of neural activity during perceptual decision-making."

22. Christian K. Machens, Ranulfo Romo, and Carlos D. Brody, "Functional, but not anatomical, separation of 'what' and 'when' in prefrontal cortex," *Journal of Neuroscience* 30 (2010): 350–60.

23. David Raposo, Matthew T. Kaufman, and Anne K. Churchland, "A category-free neural population supports evolving demands during decision-making," *Nature Neuroscience* 17 (2014): 1784–92.

CHAPTER 7. THE MEANING OF SPIKE

1. A fascinating early account of this "neural coding" debate is Perkel and Bullock's report on a Neurosciences Research Program meeting in 1968 to sort out the problem. Instead of sorting out the problem, they came away with a list of fifteen uniquely different ideas of how single neurons could send messages using spikes, one counting and fourteen timing. Donald H. Perkel and Theodore H. Bullock, "Neural coding," *NRP Bulletin* 6 (1968): 221–48. For the modern perspective on neural coding, the essential starting point is the classic book *Spikes: Exploring the Neural Code* by Fred Rieke, David Warland, Rob de Ruyter van Stevninck, and William Bialek, MIT Press, 1997.

2. For a classic example of single neuron tuning to movement direction, see Apostolos P. Georgopoulos, John F. Kalaska, Roberto Caminiti, and Joe T. Massey, "On the relations between the direction of two-dimensional arm movements and cell discharge in primate motor cortex," *Journal of Neuroscience* 2 (1982): 1527–37. A modern, detailed account is Apostolos P. Georgopoulos, Hugo Merchant, Thomas Naselaris, and Bagrat Amirikian, "Mapping of the preferred direction in the motor cortex," *Proceedings of the National Academy of Sciences USA* 104 (2007): 11068–72.

3. Nicholas G. Hatsopoulos, "Encoding in the motor cortex: Was Evarts right after all?" *Journal of Neurophysiology* 94 (2005): 2261–62.

4. J. O'Keefe and J. Dostrovsky, "The hippocampus as a spatial map: Preliminary evidence from unit activity in the freely-moving rat," *Brain Research* 34 (1971): 171–75. J. O'Keefe and D. H. Conway, "Hippocampal place units in the freely moving rat: Why they fire where they fire," *Experimental Brain Research* 31 (1978): 573–90.

5. For a review of all these types of spatial-coding cells in the hippocampus and surrounding regions, see Tom Hartley, Colin Lever, Neil Burgess, and John O'Keefe, "Space in the brain: How the hippocampal formation supports spatial cognition," *Philosophical Transactions of the Royal Society of London: Series B, Biological Sciences* 369 (2014): https://royalsocietypublishing .org/doi/full/10.1098/rstb.2012.0510.

6. C. E. Carr and M. Konishi, "A circuit for detection of interaural time differences in the brain stem of the barn owl," *Journal of Neuroscience* 10 (1990): 3227–46.

7. Michael R. DeWeese, Michael Wehr, and Anthony M. Zador, "Binary spiking in auditory cortex," *Journal of Neuroscience* 23 (2003): 7940–49.

8. Michael J. Berry, David K. Warland, and Markus Meister, "The structure and precision of retinal spike trains," *Proceedings of the National Academy of Sciences USA* 94 (1997): 5411–16.

9. Tim Gollisch and Markus Meister, "Rapid neural coding in the retina with relative spike latencies," *Science* 319 (2008): 1108–11.

10. Zachary F. Mainen and Terrence J Sejnowski, "Reliability of spike timing in neocortical neurons," *Science* 268 (1995): 1503–6.

11. Wyeth Bair and Christof Koch, "Temporal precision of spike trains in extrastriate cortex of the behaving macaque monkey," *Neural Computation* 8 (1996): 1185–202. For a more detailed analysis of spike time precision in area MT, see Giedrius T. Buračas, Anthony M. Zador, Michael R. DeWeese, and Thomas D. Albright, "Efficient discrimination of temporal patterns by motion-sensitive neurons in primate visual cortex," *Neuron* 20 (1998): 959–69.

12. The general argument is that networks of cortical neurons are too sensitive to small changes in their conditions; for a neuron to repeat a sequence of spikes with high precision requires that the rest of the network has had no change in its conditions—which is highly unlikely. This is argued most clearly in Michael London, Arnd Roth, Lisa Beeren, Michael Häusser, and Peter E. Latham, "Sensitivity to perturbations in vivo implies high noise and suggests rate coding in cortex," *Nature* 466 (2010): 123–27. See also Arunava Banerjee, Peggy Seriès, and Alexandre Pouget, "Dynamical constraints on using precise spike timing to compute in recurrent cortical networks," *Neural Computation* 20 (2008): 974–93.

13. Eugene M. Izhikevich and Gerald M. Edelman, "Large-scale model of mammalian thalamocortical systems," *Proceedings of the National Academy of Sciences USA* 105 (2008): 3593–98.

14. The models are linear-nonlinear Poisson models (LNPs) or generalized linear models (GLMs). GLMs in particular are widely used because they can be made much more complicated: one can transform the inputs to the model; include spikes from other neurons as inputs; and even include a neuron's own spikes in the past as an input, to capture how a neuron's output depends on what it did before (like the refractory period of silence after each spike). Papers laying out the GLM ideas include:

Wilson Truccolo, Uri T. Eden, Matthew R. Fellows, John P. Donoghue, and Emery N. Brown, "A point process framework for relating neural spiking activity to spiking history, neural ensemble, and extrinsic covariate effects," *Journal of Neurophysiology* 93 (2005): 1074–89.

Liam Paninski, Jonathan Pillow, and Jeremy Lewi, "Statistical models for neural encoding, decoding, and optimal stimulus design," *Progress in Brain Research* 165 (2007): 493–507.

15. Jonathan W. Pillow, Jonathon Shlens, Liam Paninski, Alexander Sher, Alan M. Litke, E. J. Chichilnisky, and Eero P. Simoncelli, "Spatio-temporal correlations and visual signalling in a complete neuronal population," *Nature* 454 (2008): 995–99.

16. Michael R. Bale, Kyle Davies, Oliver J. Freeman, Robin A. A. Ince, and Rasmus S. Petersen, "Low-dimensional sensory feature representation by trigeminal primary afferents," *Journal of Neuroscience* 33 (2013): 12003–12; Dario Campagner, Mathew H. Evans, Michael R. Bale, Andrew Erskine, and Rasmus S. Petersen, "Prediction of primary somatosensory neuron activity during active tactile exploration," *eLife* 5 (2016): https://elifesciences.org/articles/10696.

17. It's hard to find papers reporting clear examples of failures to fit predictive models, because it can be hard to know if the failure was due to the model being wrong (which is interesting), or whether it was just not implemented correctly (which is not interesting). And because people tend not to report when they failed to do something. But there are published examples. One example is Lindsay and colleagues reporting that location and movement are poor predictors of spiking in prefrontal cortex. And perhaps this is not surprising: Heitman and colleagues report that even for retinal ganglion cells, these predictive models can sometimes do a poor job of predicting spikes to natural images.

Adrian J. Lindsay, Barak F. Caracheo, Jamie J. S. Grewal, Daniel Leibovitz, and Jeremy K. Seamans, "How much does movement and location encoding impact prefrontal cortex activity? An algorithmic decoding approach in freely moving rats," *eNeuro* 5 (2018): ENEURO.0023-18.2018.

Alexander Heitman, Nora Brackbill, Martin Greschner, Alexander Sher, Alan M. Litke, and E. J. Chichilnisky, "Testing pseudo-linear models of responses to natural scenes in primate retina" (2016): https://www.biorxiv.org/content/10.1101/045336v2.

18. Admittedly, these often poor predictions of many neurons in layer two/three of the first whisker bit of cortex are made using calcium imaging (Peron et al.). Single-neuron recordings in layer four of that bit of cortex show a precisely timed spike code when a whisker hits an object (Hires et al.). So it's unclear whether the poor predictions in layer two/three are due to a coarser, slower recording of neuron activity, or due to the further processing by the local circuits within layers four and two/three and in the connections from layer four to two/three.

Simon P. Peron, Jeremy Freeman, Vijay Iyer, Caiying Guo, and Karel Svoboda, "A cellular resolution map of barrel cortex activity during tactile behavior," *Neuron* 86 (2015): 783–99.

Samuel Andrew Hires, Diego A. Gutnisky, Jianing Yu, Daniel H. O'Connor, and Karel Svoboda, "Low-noise encoding of active touch by layer 4 in the somatosensory cortex," *eLife* 4 (2015): e06619.

19. Mark Laubach, Marcelo S. Caetano, and Nandakumar S. Narayanan, "Mistakes were made: Neural mechanisms for the adaptive control of action initiation by the medial prefrontal cortex," *Journal of Physiology Paris* 109 (2015): 104–17; Mehdi Khamassi, René Quilodran, Pierre Enel, Peter F. Dominey, and Emmanuel Procyk, "Behavioral regulation and the modulation of information coding in the lateral prefrontal and cingulate cortex," *Cerebral Cortex* 25 (2015): 3197–218.

20. Perkel and Bullock's 1968 account of the Neurosciences Research Program meeting on "Neural Coding" also identified four unique "ensemble" codes, ideas for how more than one neuron could combine their spikes to send messages. These were written as aspirational ideas, as the possibility of recording ten or more neurons at the same time seemed remote.

21. For a review of how population decoding works, see Rodrigo Quian Quiroga and Stefano Panzeri, "Extracting information from neuronal populations: Information theory and decoding approaches," *Nature Reviews Neuroscience* 10 (2009): 173–85. Also good, and covering further issues in population decoding, is Andrew J. Pruszynski and Joel Zylberberg, "The language of the brain: Real-world neural population codes," *Current Opinion in Neurobiology* 58 (2019): 30–36.

22. Philipp Berens, Alexander S. Ecker, R. James Cotton, Wei Ji Ma, Matthias Bethge, and Andreas S. Tolias, "A fast and simple population code for orientation in primate V1," *Journal of Neuroscience* 32 (2012): 10618–26.

23. Joel Zylberberg, "The role of untuned neurons in sensory information coding," *bioRxiv* (2018): https://www.biorxiv.org/content/10.1101/134379v6.

24. Houman Safaai, Moritz von Heimendahl, Jose M. Sorando, Mathew E. Diamond, and Miguel Maravall, "Coordinated population activity underlying texture discrimination in rat barrel cortex," *Journal of Neuroscience* 33 (2013): 5843–55.

25. Mattia Rigotti, Omri Barak, Melissa R. Warden, Xiao-Jing Wang, Nathaniel D. Daw, Earl K. Miller, and Stefano Fusi, "The importance of mixed selectivity in complex cognitive tasks," *Nature* 49 (2013): 585–90.

Technically, Rigotti and colleagues used "pseudo-population" decoding. Neurons were recorded individually. But the task had a fixed set of events at fixed times: a cue for what the current goal was, then the first picture, then the second picture. So all the individual neurons could be aligned to the events and a population created from them.

26. Matthew L. Leavitt, Florian Pieper, Adam J. Sachs, and Julio C. Martinez-Trujillo, "Correlated variability modifies working memory fidelity in primate prefrontal neuronal ensembles," *Proceedings of the National Academy of Sciences of the USA* 114 (2017): E2494–E2503.

27. Silvia Maggi and Mark D. Humphries, "Independent population coding of the past and the present in prefrontal cortex during learning," *bioRxiv* (2020): www.biorxiv.org/content/10.1101/668962v2.

28. David Raposo, Matthew T. Kaufman, and Anne K. Churchland, "A category-free neural population supports evolving demands during decision-making," *Nature Neuroscience* 17 (2014): 1784–92.

29. There is also the inverse decoding fallacy: decoding thing *X* or *Y* is also not evidence that *X* or *Y* are all the legion of neurons knows about. Indeed, we can only check for decoding the features of the world that we train a model to learn the patterns of spikes for. There is little doubt that the legion does much more than we ask of it.

30. T. J. Brozoski, R. M. Brown, H. E. Rosvold, and P. S. Goldman, "Cognitive deficit caused by regional depletion of dopamine in prefrontal cortex of rhesus monkey," *Science* 205 (1979): 929–32. Here the lab of Patricia Goldman (later Goldman-Rakic) showed that removing big chunks of prefrontal cortex will completely prevent holding an item in memory for more than a second or two, and that removing dopamine from the prefrontal cortex has the same effect.

31. For a classic review of prefrontal cortex activity during working memory tasks, see Patricia S. Goldman-Rakic, "Cellular basis of working memory," *Neuron* 14 (1995): 477–85. An

updated view is in Earl K. Miller, Mikael Lundqvist, and André M. Bastos, "Working memory 2.0," *Neuron* 100 (2018): 463–75.

32. S. Funahashi, C. J. Bruce, and P. S. Goldman-Rakic, "Mnemonic coding of visual space in the monkey's dorsolateral prefrontal cortex," *Journal of Neurophysiology* 61 (1989): 331–49.

33. Carlos D. Brody, Adrián Hernández, Antonio Zainos, and Ranulfo Romo, "Timing and neural encoding of somatosensory parametric working memory in macaque prefrontal cortex," *Cerebral Cortex* 13 (2003): 1196–207.

34. Indeed, there is a lively debate at the time of writing about how the buffered memory is encoded by individual neurons in the prefrontal cortex. We can see that individual neurons have persistent spiking during a memory period, but we see this by averaging their activity over many repeats of the same task. The question is: does an individual neuron show persistent spiking on every trial of the task? Constantindis and colleagues say yes; Lundqvist and colleagues say no. Instead, they argue that on each trial a set of prefrontal cortex neurons show sustained spiking across the time needed to buffer memory, but not the same neurons on every trial. Which means: the memory is encoded by the population. Which is exactly what we're about to see.

Christos Constantinidis, Shintaro Funahashi, Daeyeol Lee, John D. Murray, Xue-Lian Qi, Min Wang, and Amy F. T. Arnsten, "Persistent spiking activity underlies working memory," *Journal of Neuroscience* 38 (2018): 7020–28.

Mikael Lundqvist, Pawel Herman, and Earl K. Miller, "Working memory: Delay activity, yes! Persistent activity? Maybe not," *Journal of Neuroscience* 38 (2018): 7013–19.

35. Christian K. Machens, Ranulfo Romo, and Carlos D. Brody, "Functional, but not anatomical, separation of 'what' and 'when' in prefrontal cortex," *Journal of Neuroscience* 30 (2010): 350–60. If you want a workout for your brain, also check out Christian Machens and team showing vibration-decoding and four other major results on population decoding in prefrontal cortex in one paper: Dmitry Kobak, Wieland Brendel, Christos Constantinidis, Claudia E. Feierstein, Adam Kepecs, Zachary F. Mainen, Xue-Lian Qi, Ranulfo Romo, Naoshige Uchida, and Christian K. Machens, "Demixed principal component analysis of neural population data," *eLife* 5 (2016): e10989.

36. Matthew L. Leavitt, Florian Pieper, Adam J. Sachs, and Julio C. Martinez-Trujillo, "Correlated variability modifies working memory fidelity in primate prefrontal neuronal ensembles," *Proceedings of the National Academy of Sciences of the USA* 114 (2017): E2494–E2503.

37. Silvia Maggi, Adrien Peyrache, and Mark D. Humphries, "An ensemble code in medial prefrontal cortex links prior events to outcomes during learning," *Nature Communications* 9 (2018): 2204.

38. Not all decisions are made in these regions of the cortex. As we saw in chapter 5, even simple decisions to flee are based on accumulating evidence deep in the rapid-processing center of the midbrain. Even faster decisions—like snatching your hand away from a volcanically hot roasting dish—are handled entirely by your spinal cord and brain stem.

39. K. H. Britten, W. T. Newsome, M. N. Shadlen, S. Celebrini, and J. A. Movshon, "A relationship between behavioral choice and the visual responses of neurons in macaque MT," *Visual Neuroscience* 13 (1996): 87–100.

40. Michael Shadlen's lab produced the key studies of evidence accumulation in dorsolateral prefrontal cortex (Kim and Shadlen) and parietal cortex (e.g., Roitman and Shadlen). From the same lab, an interesting direct test of formal evidence accumulation theories in parietal cortex is Kira and colleagues. Hanks and team show evidence accumulation in rats making decisions,

in both their equivalents of prefrontal and parietal cortices. Much of this work is nicely summarized in Hanks and Summerfield.

Jong-Nam Kim and Michael N. Shadlen, "Neural correlates of a decision in the dorsolateral prefrontal cortex of the macaque," *Nature Neuroscience* 2 (1999): 176–85.

Jamie D. Roitman, and Michael N. Shadlen, "Response of neurons in the lateral intraparietal area during a combined visual discrimination reaction time task," *Journal of Neuroscience* 22 (2002): 9475–89.

Shinichiro Kira, Tianming Yang, and Michael N. Shadlen, "A neural implementation of Wald's sequential probability ratio test," *Neuron* 85 (2015): 861–73.

Timothy D. Hanks, Charles D. Kopec, Bingni W. Brunton, Chunyu A. Duan, Jeffrey C. Erlich, and Carlos D. Brody, "Distinct relationships of parietal and prefrontal cortices to evidence accumulation," *Nature* 520 (2015): 220–23.

Timothy D. Hanks and Christopher Summerfield, "Perceptual decision making in rodents, monkeys, and humans," *Neuron* 93 (2017): 15–31.

41. Jamie D. Roitman and Michael N. Shadlen, "Response of neurons in the lateral intraparietal area during a combined visual discrimination reaction time task," *Journal of Neuroscience* 22 (2002): 9475–89.

42. Jochen Ditterich, Mark E. Mazurek, and Michael N. Shadlen, "Microstimulation of visual cortex affects the speed of perceptual decisions," *Nature Neuroscience* 6 (2003): 891–98.

43. *In monkeys*: Leor N. Katz, Jacob L. Yates, Jonathan W. Pillow, and Alexander C. Huk, "Dissociated functional significance of decision-related activity in the primate dorsal stream," *Nature* 35 (2016): 285–88. *In rats*: Jeffrey C. Erlich, Bingni W. Brunton, Chunyu A. Duan, Timothy D. Hanks, and Carlos D. Brody, "Distinct effects of prefrontal and parietal cortex inactivations on an accumulation of evidence task in the rat," *eLife* 4 (2015): e05457.

44. Evidence accumulation below the cortex is clearest in the striatum (Ding and Gold, "Caudate"), which we shall meet in the next chapter. Indeed, there is recent evidence that the striatum may be causally necessary for making a decision based on incoming sensory evidence (Ding and Gold, "Separate"; Yartsev et al.).

Long Ding and Joshua I. Gold, "Caudate encodes multiple computations for perceptual decisions," *Journal of Neuroscience* 30 (2010): 15747–59.

Long Ding and Joshua I. Gold, "Separate, causal roles of the caudate in saccadic choice and execution in a perceptual decision task," *Neuron* 75 (2012): 865–74.

Michael M. Yartsev, Timothy D. Hanks, Alice Misun Yoon, and Carlos D. Brody, "Causal contribution and dynamical encoding in the striatum during evidence accumulation," *eLife* 7 (2018): 34929.

45. Further evidence of the degeneracy comes from stimulation studies. Unlike the dramatic effects of stimulating area MT, stimulating in the key part of parietal cortex (lateral intraparietal cortex) only weakly alters the decision, suggesting that region had little unique causal control over the decision. See Timothy D. Hanks, Jochen Ditterich, and Michael N. Shadlen, "Microstimulation of macaque area LIP affects decision-making in a motion discrimination task," *Nature Neuroscience* 9 (2006): 682–89.

46. Miriam L. R. Meister, Jay A. Hennig, and Alexander C. Huk, "Signal multiplexing and single-neuron computations in lateral intraparietal area during decision-making," *Journal of Neuroscience* 33 (2013): 2254–67.

47. Il Memming Park, Miriam L. R. Meister, Alexander C. Huk, and Jonathan W. Pillow, "Encoding and decoding in parietal cortex during sensorimotor decision-making," *Nature Neuroscience* 17 (2014): 1395–403.

48. Roozbeh Kiani, Christopher J. Cuev, John B. Reppas, and William T. Newsome, "Dynamics of neural population responses in prefrontal cortex indicate changes of mind on single trials," *Current Biology* 24 (2014): 1542–47. For a detailed population decoding approach, see also Park et al., "Encoding and decoding in parietal cortex during sensorimotor decision-making."

49. How might the legion decide—and do it without counting? One way is for different neurons within the legion representing, say, "moving left" to jump from low to high activity at different moments while viewing the dots. Then the total count of spikes across that population of neurons is evidence for that choice (models of this are Okamoto et al.; Martí et al.). There is some evidence for this jumping from low to high spiking in individual neurons from the parietal cortex (Latimer et al.), though this claim is not without controversy (Zylberberg and Shadlen).

Hiroshi Okamoto, Yoshikazu Isomura, Masahiko Takada, and Tomoki Fukai, "Temporal integration by stochastic recurrent network dynamics with bimodal neurons," *Journal of Neurophysiology* 97 (2007): 3859–67.

Daniel Martí, Gustavo Deco, Maurizio Mattia, Guido Gigante, and Paolo Del Giudice, "A fluctuation-driven mechanism for slow decision processes in reverberant networks," *PLoS ONE* 3 (2008): e2534.

Kenneth W. Latimer, Jacob L. Yates, Miriam L. R. Meister, Alexander C. Huk, and Jonathan W. Pillow, "Single-trial spike trains in parietal cortex reveal discrete steps during decision-making," *Science* 349 (2015): 184–87. Among the responses to this paper, see Ariel Zylberberg and Michael N. Shadlen, "Cause for pause before leaping to conclusions about stepping," *bioRxiv* (2016): DOI: 10.1101/085886. The latest update to this debate is David M. Zoltowski, Kenneth W. Latimer, Jacob L. Yates, Alexander C. Huk, and Jonathan W. Pillow, "Discrete stepping and nonlinear ramping dynamics underlie spiking responses of LIP neurons during decision-making," *Neuron* 102 (2019): 1249–58.

CHAPTER 8. A MOVING EXPERIENCE

1. A model of the entire see-to-grasp pathway is in Michaels and colleagues. They show how the neural activity in each of the parietal, premotor, and motor cortex regions can be captured by a model that (a) receives as input visual information from a model of Highway Do (the dorsal stream) and (b) is trained to reproduce the velocities of fifty muscles during the reach-and-grasp. And nothing else: that alone is enough for the model to not only reproduce much of the recorded neural activity in all three regions but also to predict the neural activity that will occur during reaches to objects the model was not trained on. See Jonathan A. Michaels, Stefan Schaffelhofer, Andres Agudelo-Toro, and Hansjörg Scherberger, "A neural network model of flexible grasp movement generation," *bioRxiv* (2019): www.biorxiv.org/content/10.1101/742189v1.

2. Rodrigo Quian Quiroga, Lawrence H. Snyder, Aaron P. Batista, He Cui, and Richard A. Andersen, "Movement intention is better predicted than attention in the posterior parietal cortex," *Journal of Neuroscience* 26 (2006): 3615–20; Richard A. Andersen and He Cui, "Intention,

action planning, and decision making in parietal-frontal circuits," *Neuron* 63 (2009): 568–83; Michaels et al., "A neural network model of flexible grasp movement generation."

3. Krishna V. Shenoy, Maneesh Sahani, and Mark M. Churchland, "Cortical control of arm movements: A dynamical systems perspective," *Annual Review of Neuroscience* 36 (2013): 337–59.

4. Mark M. Churchland, John P. Cunningham, Matthew T. Kaufman, Stephen I. Ryu, and Krishna V. Shenoy, "Cortical preparatory activity: Representation of movement or first cog in a dynamical machine?" *Neuron* 68 (2010): 387–400.

5. Churchland et al., "Cortical preparatory activity: Representation of movement or first cog in a dynamical machine?"; Mark M. Churchland, Byron M. Yu, John P. Cunningham, et al., "Stimulus onset quenches neural variability: A widespread cortical phenomenon," *Nature Neuroscience* 13 (2010): 369–78.

6. Matthew T. Kaufman, Mark M. Churchland, Stephen I. Ryu, and Krishna V. Shenoy, "Cortical activity in the null space: Permitting preparation without movement," *Nature Neuroscience* 17 (2014): 440–48.

7. Sergey D. Stavisky, Jonathan C. Kao, Stephen I. Ryu, and Krishna V. Shenoy, "Motor cortical visuomotor feedback activity is initially isolated from downstream targets in output-null neural state space dimensions," *Neuron* 95 (2017): 195–208.

8. For reviews of the basal ganglia's key role in selecting actions, see:

Jonathan W. Mink, "The basal ganglia: Focused selection and inhibition of competing motor programs," *Progress in Neurobiology* 50 (1996): 381–425.

Peter Redgrave, Tony J. Prescott, and Kevin Gurney, "The basal ganglia: A vertebrate solution to the selection problem?" *Neuroscience* 89 (1999): 1009–23.

Mark D. Humphries and Tony J. Prescott, "The ventral basal ganglia, a selection mechanism at the crossroads of space, strategy, and reward," *Progress in Neurobiology* 90 (2010): 385–417.

Mark D. Humphries, "Basal ganglia: Mechanisms for action selection," in *Encyclopedia of Computational Neuroscience*, ed. D. Jaeger and R. Jung, Springer, 2014, 1–7.

9. A. J. McGeorge and R. L. Faull, "The organization of the projection from the cerebral cortex to the striatum in the rat," *Neuroscience* 29 (1989): 503–37; Nicholas R. Wall, Mauricio De La Parra, Edward M. Callaway, and Anatol C. Kreitzer, "Differential innervation of direct- and indirect-pathway striatal projection neurons," *Neuron* 79 (2013): 347–60; Barbara J. Hunnicutt, Bart C. Jongbloets, William T. Birdsong, Katrina J. Gertz, Haining Zhong, and Tianyi Mao, "A comprehensive excitatory input map of the striatum reveals novel functional organization," *eLife* 5 (2016): e19103.

10. Garrett E. Alexander and Mahlon R. DeLong, "Microstimulation of the primate neostriatum, I: Physiological properties of striatal microexcitable zones," *Journal of Neurophysiology* 53 (1985): 1401–16.

11. Petr Znamenskiy and Anthony M. Zador, "Corticostriatal neurons in auditory cortex drive decisions during auditory discrimination," *Nature* 497 (2013): 482–85; Qiaojie Xiong, Petr Znamenskiy, and Anthony M. Zador, "Selective corticostriatal plasticity during acquisition of an auditory discrimination task," *Nature* 521 (2015): 348–51.

12. Michael M. Yartsev, Timothy D. Hanks, Alice Misun Yoon, and Carlos D. Brody, "Causal contribution and dynamical encoding in the striatum during evidence accumulation," *eLife* 7 (2018): https://elifesciences.org/articles/34929; Y. Kate Hong, Clay O. Lacefield, Chris C. Rodgers, and Randy M. Bruno, "Sensation, movement, and learning in the absence of barrel cortex," *Nature* 561 (2018): 542–46.

13. Key papers on the specific effects of stimulating direct or indirect pathways of the striatum include:

Alexxai V. Kravitz, Benjamin S. Freeze, Philip R. L. Parker, Kenneth Kay, Myo T. Thwin, Karl Deisseroth, and Anatol C. Kreitzer, "Regulation of parkinsonian motor behaviours by optogenetic control of basal ganglia circuitry," *Nature* 466 (2010): 622–26.

Fatuel Tecuapetla, Sara Matias, Guillaume P. Dugue, Zachary F. Mainen, and Rui M. Costa, "Balanced activity in basal ganglia projection pathways is critical for contraversive movements," *Nature Communications* 5 (2014): 4315.

Fatuel Tecuapetla, Xin Jin, Susana Q. Lima, and Rui M. Costa, "Complementary contributions of striatal projection pathways to action initiation and execution," *Cell* 166 (2016): 703–15.

Claire E. Geddes, Hao Li, and Xin Jin, "Optogenetic editing reveals the hierarchical organization of learned action sequences," *Cell* 174 (2018): 32–43.e15.

14. Our estimates of the inputs needed to drive one principle striatal neuron to spike are deep within Mark D. Humphries, Ric Wood, and Kevin Gurney, "Dopamine-modulated dynamic cell assemblies generated by the GABAergic striatal microcircuit," *Neural Networks* 22 (2009): 1174–88. These estimates build on crucial data in Kim T. Blackwell, Uwe Czubayko, and Dietmar Plenz, "Quantitative estimate of synaptic inputs to striatal neurons during up and down states in vitro," *Journal of Neuroscience* 23 (2003): 9123–32.

15. That the principle neuron of the striatum seems designed to be choosy about what it responds to is best appreciated from models of this neuron, including:

John A. Wolf, Jason T. Moyer, Maciej T. Lazarewicz, Diego Contreras, Marianne Benoit-Marand, Patricio O'Donnell, and Leif H. Finkel, "NMDA/AMPA ratio impacts state transitions and entrainment to oscillations in a computational model of the nucleus accumbens medium spiny projection neuron," *Journal of Neuroscience* 25 (2005): 9080–95.

Jason T. Moyer, John A. Wolf, and Leif H. Finkel, "Effects of dopaminergic modulation on the integrative properties of the ventral striatal medium spiny neuron," *Journal of Neurophysiology* 98 (2007): 3731–48.

Mark D. Humphries, Nathan Lepora, Ric Wood, and Kevin Gurney, "Capturing dopaminergic modulation and bimodal membrane behaviour of striatal medium spiny neurons in accurate, reduced models," *Frontiers in Computational Neuroscience* 3 (2009): 26.

16. The output neurons of the basal ganglia are contained with the substantia nigra pars reticulata and the internal segment of the globus pallidus (or globus pallidus pars interna). In rodents, this latter structure is called the entopeduncular nucleus. Do you see now why I just called them "the output neurons"?

17. J. M. Deniau and G. Chevalier, "The lamellar organization of the rat substantia nigra pars reticulata: Distribution of projection neurons," *Neuroscience* 46 (1992): 361–77.

18. O. Hikosaka and R. H. Wurtz, "Visual and oculomotor functions of monkey substantia nigra pars reticulata, IV: Relation of substantia nigra to superior colliculus," *Journal of Neurophysiology* 49 (1983): 1285–301.

19. Thomas K. Roseberry, Moses Lee, Arnaud L. Lalive, Linda Wilbrecht, Antonello Bonci, and Anatol C. Kreitzer, "Cell-type-specific control of brainstem locomotor circuits by basal ganglia," *Cell* 164 (2016): 526–37; V. Caggiano, R. Leiras, H. Goñi-Erro, D. Masini, C. Bellardita,

J. Bouvier, V. Caldeira, G. Fisone, and O. Kiehn, "Midbrain circuits that set locomotor speed and gait selection," *Nature* 553 (2018): 455–60.

20. K. Takakusaki, T. Habaguchi, J. Ohtinata-Sugimoto, K. Saitoh, and T. Sakamoto, "Basal ganglia efferents to the brainstem centers controlling postural muscle tone and locomotion: A new concept for understanding motor disorders in basal ganglia dysfunction," *Neuroscience* 119 (2003): 293–308.

21. F. A. Middleton and P. L. Strick, "Basal-ganglia 'projections' to the prefrontal cortex of the primate," *Cerebral Cortex* 12 (2002): 926–35; Ágnes L. Bodor, Kristóf Giber, Zita Rovó, István Ulbert, and László Acsády, "Structural correlates of efficient GABAergic transmission in the basal ganglia-thalamus pathway," *Journal of Neuroscience* 28 (2008): 3090–102.

22. G. Chevalier and J. M. Deniau, "Disinhibition as a basic process in the expression of striatal function," *Trends in Neurosciences* 13 (1990): 277–80; Jeremy R. Edgerton and Dieter Jaeger, "Optogenetic activation of nigral inhibitory inputs to motor thalamus in the mouse reveals classic inhibition with little potential for rebound activation," *Frontiers in Cellular Neuroscience* 8 (2014): 36.

23. O. Hikosaka and R. H. Wurtz, "Modification of saccadic eye movements by GABA-related substances, II: Effects of muscimol in monkey substantia nigra pars reticulata," *Journal of Neurophysiology* 53 (1985): 292–308.

24. Arthur Leblois, Wassilios Meissner, Erwan Bezard, Bernard Bioulac, Christian E. Gross, and Thomas Boraud, "Temporal and spatial alterations in GPi neuronal encoding might contribute to slow down movement in Parkinsonian monkeys," *European Journal of Neuroscience* 24 (2006): 1201–8; Mark D. Humphries, Robert D. Stewart, and Kevin N. Gurney, "A physiologically plausible model of action selection and oscillatory activity in the basal ganglia," *Journal of Neuroscience* 26 (2006): 12921–42.

25. Dorothy E. Oorschot, "Total number of neurons in the neostriatal, pallidal, subthalamic, and substantia nigral nuclei of the rat basal ganglia: A stereological study using the cavalieri and optical disector methods," *Journal of Comparative Neurology* 366 (1996): 580–99.

26. How exactly the basal ganglia's circuit selects and switches between actions is a fascinating but frankly mind-boggling topic. Others and I have built many detailed models of how exactly the direct and indirect pathways from the striatum compete and what the other nuclei of the basal ganglia contribute. For an overview, start with Mark D. Humphries, "Basal ganglia: Mechanisms for action selection," in *Encyclopedia of Computational Neuroscience*, ed. D. Jaeger and R. Jung, Springer, 2014, 1–7. For major models, see Kevin Gurney, Tony J. Prescott, and Peter Redgrave, "A computational model of action selection in the basal ganglia I: A new functional anatomy," *Biological Cybernetics* 85 (2001): 401–10 (see also Part II, on pages 411–23 of the same issue); Mark D. Humphries, Robert D. Stewart, and Kevin N. Gurney, "A physiologically plausible model of action selection and oscillatory activity in the basal ganglia," *Journal of Neuroscience* 26 (2006): 12921–42; and Michael J. Frank, "Dynamic dopamine modulation in the basal ganglia: A neurocomputational account of cognitive deficits in medicated and nonmedicated Parkinsonism," *Journal of Cognitive Neuroscience* 17 (2005): 51–72.

27. Apostolos P. Georgopoulos, Andrew B. Schwartz, and Ronald E. Kettner, "Neuronal population coding of movement direction," *Science* 233 (1986): 1416–19.

28. The decoding of arm-movement direction from tuned neurons in motor cortex was the first example of "vector" coding. Take a set of neurons that each have a preferred direction of movement—some send the most spikes when the arm moves diagonally up and to the right, some when the arm moves down and a little to the left, and so on. Vector coding works by averaging across these neurons. First, for the current arm movement, give each neuron a weight according to the number of spikes it is sending. Then take the weighted average of the directions preferred by these neurons (high weights mean greater contribution to the average). The average direction turns out to be very close to the actual direction of arm movement.

29. Jose M. Carmena, Mikhail A. Lebedev, Roy E. Crist, Joseph E. O'Doherty, David M. Santucci, Dragan F. Dimitrov, Parag G. Patil, Craig S. Henriquez, and Miguel A. L. Nicolelis, "Learning to control a brain-machine interface for reaching and grasping by primates," *PLoS Biology* 1 (2003): E42; Jose M. Carmena, Mikhail A. Lebedev, Craig S. Henriquez, and Miguel A. L. Nicolelis, "Stable ensemble performance with single-neuron variability during reaching movements in primates," *Journal of Neuroscience* 25 (2005): 10712–16.

30. Stefan Schaffelhofer, Andres Agudelo-Toro, and Hansjörg Scherberger, "Decoding a wide range of hand configurations from macaque motor, premotor, and parietal cortices," *Journal of Neuroscience* 35 (2015): 1068–81.

31. Mark M. Churchland, John P. Cunningham, Matthew T. Kaufman, Justin D. Foster, Paul Nuyujukian, Stephen I. Ryu, and Krishna V. Shenoy, "Neural population dynamics during reaching," *Nature* 487 (2012): 51–56.

32. Abigail A. Russo, Sean R. Bittner, Sean M. Perkins, Jeffrey S. Seely, Brian M. London, Antonio H. Lara, Andrew Miri, Najja J. Marshall, Adam Kohn, Thomas M. Jessell, Laurence F. Abbott, John P. Cunningham, and Mark M. Churchland, "Motor cortex embeds muscle-like commands in an untangled population response," *Neuron* 97 (2018): 953–66.

33. Russo et al., "Motor cortex embeds muscle-like commands in an untangled population response."

34. Chethan Pandarinath, Daniel J. O'Shea, Jasmine Collins, Rafal Jozefowicz, Sergey D. Stavisky, Jonathan C. Kao, Eric M. Trautmann, Matthew T. Kaufman, Stephen I. Ryu, Leigh R. Hochberg, Jaimie M. Henderson, Krishna V. Shenoy, L. F. Abbott, and David Sussillo, "Inferring single-trial neural population dynamics using sequential auto-encoders," *Nature Methods* 15 (2018): 805–15.

35. Juan A. Gallego, Matthew G. Perich, Stephanie N. Naufel, Christian Ethier, Sara A. Solla, and Lee E. Miller, "Cortical population activity within a preserved neural manifold underlies multiple motor behaviors," *Nature Communications* 9 (2018): 4233.

36. Maria Soledad Esposito, Paolo Capelli, and Silvia Arber, "Brainstem nucleus MdV mediates skilled forelimb motor tasks," *Nature* 508 (2014): 351–56.

37. Bror Alstermark and Tadashi Isa, "Circuits for skilled reaching and grasping," *Annual Review of Neuroscience* 35 (2012): 559–78.

38. Rune W. Berg, Aidas Alaburda, and Jorn Hounsgaard, "Balanced inhibition and excitation drive spike activity in spinal half-centers," *Science* 315 (2007): 390–93; Peter C. Petersen and Rune W. Berg, "Lognormal firing rate distribution reveals prominent fluctuation-driven regime in spinal motor networks," *eLife* 5 (2016): e18805.

39. Masaki Ueno, Yuka Nakamura, Jie Li, Zirong Gu, Jesse Niehaus, Mari Maezawa, Steven A. Crone, Martyn Goulding, Mark L. Baccei, and Yutaka Yoshida, "Corticospinal circuits

from the sensory and motor cortices differentially regulate skilled movements through distinct spinal interneurons," *Cell Reports* 23 (2018): 1286–300.

40. Roger N. Lemon, "Descending pathways in motor control," *Annual Review of Neuroscience* 31 (2008): 195–218.

CHAPTER 9. SPONTANEITY

1. Alan Peters and Bertram R. Payne, "Numerical relationships between geniculocortical afferents and pyramidal cell modules in cat primary visual cortex," *Cerebral Cortex* 3 (1993): 69–78; Bashir Ahmed, John C. Anderson, Rodney J. Douglas, Kevan A. C. Martin, and J. Charmaine Nelson, "Polyneuronal innervation of spiny stellate neurons in cat visual cortex," *Journal of Comparative Neurology* 341 (1994): 39–49.

2. For a classic, readable comprehensive review of the brain's default network, see Randy L. Buckner, Jessica R. Andrews-Hanna, and Daniel L. Schacter, "The brain's default network: Anatomy, function, and relevance to disease," *Annals of the New York Academy of Sciences* 1124 (2008): 1–38; updated in Randy L. Buckner and Lauren M. DiNicola, "The brain's default network: Updated anatomy, physiology, and evolving insights," *Nature Reviews Neuroscience* 20 (2019): 593–608. The activity in the default network also constantly waxes and wanes in synchrony across its regions. See, e.g., Michael D. Fox, Abraham Z. Snyder, Justin L. Vincent, Maurizio Corbetta, David C. Van Essen, and Marcus E. Raichle, "The human brain is intrinsically organized into dynamic, anticorrelated functional networks," *Proceedings of the National Academy of Sciences USA* 102 (2005): 9673–78.

3. This account of the firing of cortical neurons during sleep is drawn from Edward E. Evarts, "Temporal patterns of discharge of pyramidal tract neurons during sleep and waking in the monkey," *Journal of Neurophysiology* 27 (1964): 152–71; Alain Destexhe, Diego Contreras, and Mircea Steriade, "Spatiotemporal analysis of local field potentials and unit discharges in cat cerebral cortex during natural wake and sleep states," *Journal of Neuroscience* 19 (1999): 4595–608; and M. Steriade, I. Timofeev, and F. Grenier, "Natural waking and sleep states: A view from inside neocortical neurons," *Journal of Neurophysiology* 85 (2001): 1969–85.

4. Elda Arrigoni, Michael C. Chen, and Patrick M. Fuller, "The anatomical, cellular, and synaptic basis of motor atonia during rapid eye movement sleep," *Journal of Physiology* 594 (2016): 5391–414.

5. I draw here on excellent reviews of brain development that cover spontaneous activity: N. Dehorter, L. Vinay, C. Hammond, and Y. Ben-Ari, "Timing of developmental sequences in different brain structures: Physiological and pathological implications," *European Journal of Neuroscience* 35 (2012): 1846–56; Alexandra H. Leighton and Christian Lohmann, "The wiring of developing sensory circuits—From patterned spontaneous activity to synaptic plasticity mechanisms," *Frontiers in Neural Circuits* 10 (2016): 71; and Heiko J. Luhmann, Anne Sinning, Jenq-Wei Yang, Vicente Reyes-Puerta, Maik C. Stüttgen, Sergei Kirischuk, and Werner Kilb, "Spontaneous neuronal activity in developing neocortical networks: From single cells to large-scale interactions," *Frontiers in Neural Circuits* 10 (2016): 40.

6. Nir Kalisman, Gilad Silberberg, and Henry Markram, "The neocortical microcircuit as a tabula rasa," *Proceedings of the National Academy of Sciences USA* 102 (2005): 880–85.

7. Jean-Vincent Le Bé and Henry Markram, "Spontaneous and evoked synaptic rewiring in the neonatal neocortex," *Proceedings of the National Academy of Sciences USA* 103 (2006): 13214–19.

8. A classic review of self-generating spikes is: Rodolfo R. Llinas, "The intrinsic electrophysiological properties of mammalian neurons: Insights into central nervous system function," *Science* 242 (1998): 1654–64.

9. For a full account of pacemaking in the basal ganglia, see D. James Surmeier, Jeff N. Mercer, and C. Savio Chan, "Autonomous pacemakers in the basal ganglia: Who needs excitatory synapses anyway?" *Current Opinion in Neurobiology* 15 (2005): 312–18.

10. Leighton and Lohmann, "The wiring of developing sensory circuits—From patterned spontaneous activity to synaptic plasticity mechanisms"; Luhmann et al., "Spontaneous neuronal activity in developing neocortical networks: From single cells to large-scale interactions."

11. Morgane Le Bon-Jego and Rafael Yuste, "Persistently active, pacemaker-like neurons in neocortex," *Frontiers in Neuroscience* 1 (2007): 123–29.

12. Bruce P. Bean, "The action potential in mammalian central neurons," *Nature Reviews Neuroscience* 8 (2007): 451–65.

13. Bu-Qing Mao, Farid Hamzei-Sichani, Dmitriy Aronov, Robert C. Froemke, and Rafael Yuste, "Dynamics of spontaneous activity in neocortical slices," *Neuron* 32 (2001): 883–98; Rosa Cossart, Dmitriy Aronov, and Rafael Yuste, "Attractor dynamics of network UP states in the neocortex," *Nature* 423 (2003): 283–88.

14. Maria V. Sanchez-Vives and David A. McCormick, "Cellular and network mechanisms of rhythmic recurrent activity in neocortex," *Nature Neuroscience* 3 (2000): 1027–34.

And why is it important to get the correct salty water onto the slices of cortex? Because a neuron's voltage is determined by the difference between the concentration of charged ions inside and outside its skin. So for a neuron in a bit of sliced brain to behave like it would in the actual brain, the precise ingredients of the soup it sits in are crucial.

15. Takuya Sasaki, Norio Matsuki, and Yuji Ikegaya, "Metastability of active CA3 networks," *Journal of Neuroscience* 27 (2007): 517–28.

16. Counting feedback loops to one pyramidal neuron. Say we have N pyramidal neurons in the neighborhood of our source neuron, and that source neuron has a probability p of connecting to each neuron in that neighborhood. Then the expected number of loops of length k is approximately: $E[k] = p^k N^{k-1}$ (where $k = 2$ is a direct feedback loop to the source neuron). In the text, I plug in $N = 10,000$ and $p = 0.1$, as rough cortex-like numbers. This model assumes that all connections are equally likely. They are not. For one thing, neurons farther apart are less likely to be connected. For another, we've already learned on our journey that in visual cortex pyramidal neurons with similar tuning are more likely to be connected.

17. Tom Binzegger, Rodney J. Douglas, and Kevan A. C. Martin, "A quantitative map of the circuit of cat primary visual cortex," *Journal of Neuroscience* 24 (2004): 8441–53.

18. Daniel J. Felleman and David C. Van Essen, "Distributed hierarchical processing in the primate cerebral cortex," *Cerebral Cortex* 1 (1991): 1–47.

19. S. Murray Sherman, "Thalamic relays and cortical functioning," *Progress in Brain Research* 149 (2005): 107–26.

20. How the generated activity of a network of neurons depends on the fine detail of that network, some key examples:

Alfonso Renart, Rubén Moreno-Bote, Xiao-Jing Wang, and Néstor Parga, "Mean-driven and fluctuation-driven persistent activity in recurrent networks," *Neural Computation* 19 (2007): 1–46.

David Sussillo and L. F. Abbott, "Generating coherent patterns of activity from chaotic neural networks," *Neuron* 63 (2009): 544–57.

H. Francis Song, Guangyu R. Yang, and Xiao-Jing Wang, "Training excitatory-inhibitory recurrent neural networks for cognitive tasks: A simple and flexible framework," *PLoS Computational Biology* 12 (2016): e1004792.

21. Wolfgang Maass, Thomas Natschläger, and Henry Markram, "Real-time computing without stable states: A new framework for neural computation based on perturbations," *Neural Computation* 14 (2002): 2531–60.

22. Guillaume Hennequin, Tim P. Vogels, and Wulfram Gerstner, "Optimal control of transient dynamics in balanced networks supports generation of complex movements," *Neuron* 82 (2014): 1394–406; David Sussillo, Mark M. Churchland, Matthew T. Kaufman, and Krishna V. Shenoy, "A neural network that finds a naturalistic solution for the production of muscle activity," *Nature Neuroscience* 18 (2015): 1025–33; Jonathan A. Michaels, Benjamin Dann, and Hansjörg Scherberger, "Neural population dynamics during reaching are better explained by a dynamical system than representational tuning," *PLoS Computational Biology* 12 (2016): e1005175.

23. For reviews of networks that generate movement—so-called central pattern generators—see Eve Marder and Dirk Bucher, "Central pattern generators and the control of rhythmic movements," *Current Biology* 11 (2001): R986-96; Alan I. Selverston, "Invertebrate central pattern generator circuits," *Philosophical Transactions of the Royal Society of London B: Biological Science* 365 (2010): 2329–45.

24. Xiao-Jing Wang, "Synaptic basis of cortical persistent activity: The importance of NMDA receptors to working memory," *Journal of Neuroscience* 19 (1999): 9587–603; Francesca Barbieri and Nicolas Brunel, "Can attractor network models account for the statistics of firing during persistent activity in prefrontal cortex?" *Frontiers in Neuroscience* 2 (2008): 114–22.

25. Valerio Mante, David Sussillo, Krishna V. Shenoy, and William T. Newsome, "Context-dependent computation by recurrent dynamics in prefrontal cortex," *Nature* 503 (2013): 78–84.

26. Xiao-Jing Wang, "Probabilistic decision making by slow reverberation in cortical circuits," *Neuron* 36 (2002): 955–68; Kong-Fatt Wong and Xiao-Jing Wang, "A recurrent network mechanism of time integration in perceptual decisions," *Journal of Neuroscience* 26 (2006): 1314–28.

27. Javier A. Caballero, Mark D. Humphries, and Kevin N. Gurney, "A probabilistic, distributed, recursive mechanism for decision-making in the brain," *PLoS Computational Biology* 14 (2018): e1006033.

28. And this hypothesis can be extended to the dark neurons proper. Perhaps we never see them, as we do not ask the brain something that needs the self-sustaining dynamics they create.

CHAPTER 10. BUT A MOMENT IN TIME

1. Lionel G. Nowak and Jean Bullier, "The timing of information transfer in the visual system," in *Extrastriate Cortex in Primates*, ed. Kathleen S. Rockland, Jon H. Kaas, and Alan Peters, Springer, 1997, 205–41.

2. The study of how long it takes the brain to process something is called "mental chronometry." As we'll learn in this section, surprisingly deep insights into how the brain processes information can be gained by nothing more than a clever experimental design and a stopwatch. For a brief history, see Michael I. Posner, "Timing the brain: Mental chronometry as a tool in neuroscience," *PLoS Biology* 3 (2005): e51.

3. Simon Thorpe, Denis Fize, and Catherine Marlot, "Speed of processing in the human visual system," *Nature* 381 (1996): 520–22; Michèle Fabre-Thorpe, Arnaud Delorme, Catherine Marlot, and Simon Thorpe, "A limit to the speed of processing in ultra-rapid visual categorization of novel natural scenes," *Journal of Cognitive Neuroscience* 13 (2001): 171–80.

4. Terrence R. Stanford, Swetha Shankar, Dino P. Massoglia, M. Gabriela Costello, and Emilio Salinas, "Perceptual decision making in less than 30 milliseconds," *Nature Neuroscience* 13 (2010): 379–85.

5. Stanislas Dehaene, "The organization of brain activations in number comparison: Event-related potentials and the additive-factors method," *Journal of Cognitive Neuroscience* 8 (1996): 47–68.

6. Data in this section on response times in the random dot motion task come from:

Jamie D. Roitman and Michael N. Shadlen, "Response of neurons in the lateral intraparietal area during a combined visual discrimination reaction time task," *Journal of Neuroscience* 22 (2002): 9475–89.

Jon Palmer, Alexander C. Huk, and Michael N. Shadlen, "The effect of stimulus strength on the speed and accuracy of a perceptual decision," *Journal of Vision* 5 (2005): 376–404.

7. Running the numbers of possible paths from a starting neuron. Say we have N possible targets to leap across a gap onto; that the failure rate of each gap is f (say, 0.75 for 75 percent); and that there is a probability $p[spike]$ of the target neuron itself sending a spike in the next 10 milliseconds. Then the number of possible paths forward is: $N \times (1-f) \times p[spike]$. For the numbers in the text this gives:

$$7500 \times (1-0.75) \times 0.01 = 19 \text{ neurons (rounding up)}.$$

But after two leaps, each of those 19 neurons will also have 19 possible targets, so the number of possible forward paths is $19 \times 19 = 3,516$. After three leaps, $19 \times 19 \times 19$. And so on ...

8. For more on the idea that a population of neurons can be primed by spontaneous activity, see:

Bruce W. Knight, "Dynamics of encoding in a population of neurons," *Journal of General Physiology* 59 (1972): 734–66.

Wulfram Gerstner, "How can the brain be so fast?" in *23 Problems in Systems Neuroscience*, ed. J. Leo van Hemmen and Terence J. Sejnowski, Oxford University Press, 2006, 135–42.

Tatjana Tchumatchenko, Aleksey Malyshev, Fred Wolf, and Maxim Volgushev, "Ultrafast population encoding by cortical neurons," *Journal of Neuroscience* 31 (2011): 12171–79.

Maxim Volgushev, "Cortical specializations underlying fast computations," *Neuroscientist* 22 (2016): 145–64.

9. Marcus E. Raichle, "Two views of brain function," *Trends in Cognitive Sciences* 14 (2010): 180–90.

10. Colin Blakemore and Grahame F. Cooper, "Development of the brain depends on the visual environment," *Nature* 228 (1970): 477–78.

11. D. H. Hubel and T. N. Wiesel, "The period of susceptibility to the physiological effects of unilateral eye closure in kittens," *Journal of Physiology* 206 (1970): 419–36.

12. There are many variations on this theory of how the visual cortices predict information coming from the eye. Some key theory papers include:

Rajesh P. N. Rao and Dana H. Ballard, "Predictive coding in the visual cortex: A functional interpretation of some extra-classical receptive-field effects," *Nature Neuroscience* 2 (1999): 79–87.

Tai Sing Lee and David Mumford, "Hierarchical Bayesian inference in the visual cortex," *Journal of the Optical Society of America: A* 20 (2003): 1434–48.

Gergő Orbán, Pietro Berkes, József Fiser, and Máté Lengyel, "Neural variability and sampling-based probabilistic representations in the visual cortex," *Neuron* 92 (2016): 530–43.

13. Michael L. Platt and Paul W. Glimcher, "Neural correlates of decision variables in parietal cortex," *Nature* 400 (1999): 233–38.

14. Michael N. Shadlen and William T. Newsome, "Neural basis of a perceptual decision in the parietal cortex (area LIP) of the rhesus monkey," *Journal of Neurophysiology* 86 (2001): 1916–36.

15. Hans Supèr, Chris van der Togt, Henk Spekreijse, and Victor A. F. Lamme, "Internal state of monkey primary visual cortex (V1) predicts figure-ground perception," *Journal of Neuroscience* 23 (2003): 3407–14.

16. Guido Hesselmann, Christian A. Kell, Evelyn Eger, and Andreas Kleinschmidt, "Spontaneous local variations in ongoing neural activity bias perceptual decisions," *Proceedings of the National Academy of Sciences USA* 105 (2008): 10984–89.

17. Rafal Bogacz, Eric Brown, Jeff Moehlis, Philip Holmes, and Jonathan D. Cohen, "The physics of optimal decision making: A formal analysis of models of performance in two-alternative forced-choice tasks," *Psychological Review* 113 (2006): 700–765; Birte U. Forstmann, Scott Brown, Gilles Dutilh, Jane Neumann, and Eric-Jan Wagenmakers, "The neural substrate of prior information in perceptual decision making: A model-based analysis," *Frontiers in Human Neuroscience* 4 (2010): 40; Javier A. Caballero, Mark D. Humphries, and Kevin N. Gurney, "A probabilistic, distributed, recursive mechanism for decision-making in the brain," *PLoS Computational Biology* 14 (2018): e1006033.

18. Michael D. Fox, Abraham Z. Snyder, Justin L. Vincent, and Marcus E. Raichle, "Intrinsic fluctuations within cortical systems account for intertrial variability in human behavior," *Neuron* 56 (2007): 171–84.

19. Joshua I. Glaser, Matthew G. Perich, Pavan Ramkumar, Lee E. Miller, and Konrad P. Kording, "Population coding of conditional probability distributions in dorsal premotor cortex," *Nature Communications* 9 (2018): 1788.

20. József Fiser, Pietro Berkes, Gergő Orbán, and Máté Lengyel, "Statistically optimal perception and learning: From behavior to neural representations," *Trends in Cognitive Sciences* 14 (2010): 119–30.

21. Dario L. Ringach, "Spontaneous and driven cortical activity: Implications for computation," *Current Opinion in Neurobiology* 19 (2009): 439–44.

22. Amos Arieli, Alexander Sterkin, Amiram Grinvald, and Ad Aertsen, "Dynamics of ongoing activity: Explanation of the large variability in evoked cortical responses," *Science* 273 (1996): 1868–71; M. Tsodyks, T. Kenet, A. Grinvald, and A. Arieli, "Linking spontaneous activity of single cortical neurons and the underlying functional architecture," *Science* 286 (1999): 1943–46; Tal Kenet, Dmitri Bibitchkov, Misha Tsodyks, Amiram Grinvald, and Amos Arieli, "Spontaneously emerging cortical representations of visual attributes," *Nature* 425 (2003): 954–56.

23. József Fiser, Chiayu Chiu, and Michael Weliky, "Small modulation of ongoing cortical dynamics by sensory input during natural vision," *Nature* 431 (2004): 573–78.

24. Artur Luczak, Péter Barthó, and Kenneth D. Harris, "Spontaneous events outline the realm of possible sensory responses in neocortical populations," *Neuron* 62 (2009): 413–25.

25. Abhinav Singh, Adrien Peyrache, and Mark D. Humphries, "Medial prefrontal cortex population activity is plastic irrespective of learning," *Journal of Neuroscience* 39 (2019): 3470–83.

26. Pietro Berkes, Gergő Orbán, Máté Lengyel, and József Fiser, "Spontaneous cortical activity reveals hallmarks of an optimal internal model of the environment," *Science* 331 (2011): 83–87.

27. For some evidence of when the evoked and spontaneous distributions of spikes barely differ, see Adrien Wohrer, Mark D. Humphries, and Christian Machens, "Population-wide distributions of neural activity during perceptual decision-making," *Progress in Neurobiology* 103 (2013): 156–93.

28. Graham E. Budd, "Early animal evolution and the origins of nervous systems," *Philosophical Transactions of the Royal Society B: Biological Sciences* 370 (2015): https://royalsocietypublishing.org/doi/10.1098/rstb.2015.0037.

29. Travis Monk and Michael G. Paulin, "Predation and the origin of neurons," *Brain, Behavior, and Evolution* 84 (2014): 246–61.

30. Gáspár Jékely, Fred Keijzer, and Peter Godfrey-Smith, "An option space for early neural evolution," *Philosophical Transactions of the Royal Society B: Biological Sciences* 370 (2015): https://royalsocietypublishing.org/doi/10.1098/rstb.2015.0181.

31. For vastly more on what happens on the slower timescales of a brain, see Robert Sapolsky's monumental *Behave*, Vintage Books, 2017.

CODA. THE FUTURE OF SPIKES

1. Doubling time of the number of recorded neurons in 2011 is taken from Ian Stevenson and Konrad Kording, "How advances in neural recording affect data analysis," *Nature Neuroscience* 14 (2011): 139–42. The estimate at the time of writing in early 2020 is taken from Ian Stevenson's lab website: https://stevenson.lab.uconn.edu/scaling/.

2. James J. Jun, Nicholas A. Steinmetz, Joshua H. Siegle, Daniel J. Denman, Marius Bauza, Brian Barbarits, Albert K. Lee, Costas A. Anastassiou, Alexandru Andrei, Çağatay Aydın, Mladen Barbic, Timothy J. Blanche, Vincent Bonin, João Couto, Barundeb Dutta, Sergey L. Gratiy, Diego A. Gutnisky, Michael Häusser, Bill Karsh, Peter Ledochowitsch, Carolina Mora Lopez, Catalin Mitelut, Silke Musa, Michael Okun, Marius Pachitariu, Jan Putzeys, P. Dylan

Rich, Cyrille Rossant, Wei-lung Sun, Karel Svoboda, Matteo Carandini, Kenneth D. Harris, Christof Koch, John O'Keefe, and Timothy D. Harris, "Fully integrated silicon probes for high-density recording of neural activity," *Nature* 551 (2017): 232–36.

3. Nicholas A. Steinmetz, Peter Zatka-Haas, Matteo Carandini, and Kenneth D. Harris, "Distributed coding of choice, action, and engagement across the mouse brain," *Nature* 576 (2019): 266–73.

4. Valentin Braitenberg and Almut Schuz, *Cortex: Statistics and Geometry of Neuronal Connectivity*, 2nd ed., Springer, 1998.

5. At the time of writing, the largest recordings in mammals are the approximately ten thousand neurons in the mouse V1 by Carsen Stringer, Marius Pachitariu, Nicholas Steinmetz, Charu Bai Reddy, Matteo Carandini, and Kenneth D. Harris, "Spontaneous behaviors drive multidimensional, brainwide activity," *Science* 364 (2019): 255; and Carsen Stringer, Marius Pachitariu, Nicholas Steinmetz, Matteo Carandini, and Kenneth D. Harris, "High-dimensional geometry of population responses in visual cortex," *Nature* 571 (2019): 361–65. The largest recordings of all are the tens of thousands in the baby zebrafish by Misha B. Ahrens, Jennifer M. Li, Michael B. Orger, Drew N. Robson, Alexander F. Schier, Florian Engert, and Ruben Portugues, "Brain-wide neuronal dynamics during motor adaptation in zebrafish" *Nature* 485 (2012): 471–77; and Nikita Vladimirov, Yu Mu, Takashi Kawashima, Davis V. Bennett, Chao-Tsung Yang, Loren L. Looger, Philipp J. Keller, Jeremy Freeman, and Misha B. Ahrens, "Light-sheet functional imaging in fictively behaving zebrafish," *Nature Methods* 11 (2014): 883–84.

6. Peter Ledochowitsch, Lawrence Huang, Ulf Knoblich, Michael Oliver, Jerome Lecoq, Clay Reid, Lu Li, Hongkui Zeng, Christof Koch, Jack Waters, Saskia E. J. de Vries, and Michael A. Buice, "On the correspondence of electrical and optical physiology in in vivo population-scale two-photon calcium imaging," *bioRxiv* (2019): DOI: 10.1101/800102.

7. For a review of the first two decades of work on voltage imaging of neurons, see M. Zochowski, M. Wachowiak, C. X. Falk, L. B. Cohen, Y. W. Lam, S. Antic, and D. Zecevic, "Imaging membrane potential with voltage-sensitive dyes," *Biological Bulletin* 198 (2000): 1–21.

8. *Leeches*: Kevin L. Briggman, Henry D. I. Abarbanel, and William B. Kristan Jr., "Optical imaging of neuronal populations during decision-making," *Science* 307 (2005): 896–901. *Seaslugs*: Jian-young Wu, Lawrence B. Cohen, and Chun Xiao Falk, "Neuronal activity during different behaviors in Aplysia: A distributed organization?" *Science* 263 (1994): 820–23.

9. The latter half of 2019 saw four major papers on breakthroughs in voltage imaging in mammals. In publication order:

May: Y. Adam et al., "Voltage imaging and optogenetics reveal behaviour-dependent changes in hippocampal dynamics," *Nature* 569 (2019): 413–17.

August: A. S. Abdelfattah et al., "Bright and photostable chemigenetic indicators for extended in vivo voltage imaging," *Science* 365 (2019): 699–704.

October: K. D. Piatkevich et al., "Population imaging of neural activity in awake behaving mice," *Nature* 574 (2019): 413–17.

December: V. Villette et al., "Ultrafast two-photon imaging of a high-gain voltage indicator in awake behaving mice," *Cell* 179 (2019): 1590–608.

10. For an entertainingly skeptical view on whether more brain data will help us, see Yves Frégnac, "Big data and the industrialization of neuroscience: A safe roadmap for understanding the brain?" *Science* 358 (2017): 470–77.

11. A. R. Powers, C. Mathys, and P. R. Corlett, "Pavlovian conditioning-induced hallucinations result from overweighting of perceptual priors," *Science* 357 (2017): 596–600.

12. *Parkinson's disease:* Jones G. Parker, Jesse D. Marshall, Biafra Ahanonu, Yu-Wei Wu, Tony Hyun Kim, Benjamin F. Grewe, Yanping Zhang, Jin Zhong Li, Jun B. Ding, Michael D. Ehlers, and Mark J. Schnitzer, "Diametric neural ensemble dynamics in parkinsonian and dyskinetic states," *Nature* 557 (2018): 177–82. *Fragile-X syndrome:* Cian O'Donnell, J. Tiago Gonçalves, Carlos Portera-Cailliau, and Terrence J. Sejnowski, "Beyond excitation/inhibition imbalance in multidimensional models of neural circuit changes in brain disorders," *eLife* 6 (2017): e26724.

13. Elon Musk and Neuralink, "An integrated brain-machine interface platform with thousands of channels," *bioRxiv* (2019): DOI: 10.1101/703801.

14. Ryan M. Neely, David Piech, Samantha R. Santacruz, Michel M. Maharbiz, and Jose M. Carmena, "Recent advances in neural dust: Towards a neural interface platform," *Current Opinion in Neurobiology* 50 (2018): 64–71.

15. Raviv Pryluk, Yoav Kfir, Hagar Gelbard-Sagiv, Itzhak Fried, and Rony Paz, "A tradeoff in the neural code across regions and species," *Cell* 176 (2019): 597–609.

16. Sticking "neuro" in front of a word to make a new field or concept that has no factual basis in neuroscience is called "neurobollocks."

Steven Poole, "Your brain on pseudoscience: The rise of popular neurobollocks," *New Statesman* 6 (September 2012): www.newstatesman.com/culture/books/2012/09/your-brain-pseudoscience-rise-popular-neurobollocks.

For evidence that such things exist, see, for example:

Neuromarketing: Eben Harrell, "Neuromarketing: What you need to know," *Harvard Business Review* 23 (January 2019): https://hbr.org/2019/01/neuromarketing-what-you-need-to-know.

Neurolaw: Eryn Brown, "The brain, the criminal and the courts," *Knowable Magazine*, August 30, 2019, www.knowablemagazine.org/article/mind/2019/neuroscience-criminal-justice.

Neurocriticism: Rob Horning, "Neurocriticism and neurocapitalism," *PopMatters*, April 22, 2010, www.popmatters.com/neurocriticism-and-neurocapitalism-2496196908.html.

17. Mark Humphries, "How your brain learns to fear," *Medium*, June 1, 2017, https://medium.com/s/theories-of-mind/how-your-brain-learns-to-fear-a7bd3ab38ed9.

18. Joseph E. LeDoux, "Coming to terms with fear," *Proceedings of the National Academy of Sciences of the USA* 111 (2014): 2871–78; Lisa Feldman Barrett, *How Emotions Are Made*, Pan Books, 2017.

19. Benjamin F. Grewe, Jan Gründemann, Lacey J. Kitch, Jerome A. Lecoq, Jones G. Parker, Jesse D. Marshall, Margaret C. Larkin, Pablo E. Jercog, Francois Grenier, Jin Zhong Li, Andreas Lüthi, and Mark J. Schnitzer, "Neural ensemble dynamics underlying a long-term associative memory," *Nature* 543 (2017): 670–75.

20. A. Demertzi, E. Tagliazucchi, S. Dehaene, G. Deco, P. Barttfeld, F. Raimondo, C. Martial, D. Fernández-Espejo, B. Rohaut, H. U. Voss, N. D. Schiff, A. M. Owen, S. Laureys, L. Naccache, and J. D. Sitt, "Human consciousness is supported by dynamic complex patterns of brain signal coordination," *Science Advances* 5 (2019): eaat7603.

21. Giulio Tononi and Olaf Sporns, "Measuring information integration," *BMC Neuroscience* 4 (2003): 31; Giulio Tononi, "An information integration theory of consciousness," *BMC Neuroscience* 5 (2004): 42.

22. Roger Penrose, *The Emperor's New Mind*, Oxford University Press, 1989; Roger Penrose, *Shadows of the Mind*, Oxford University Press, 1994.

23. Lynne Rudder Baker, "Review of *Contemporary Dualism: A Defense*," *Notre Dame Philosophical Reviews*, October 16, 2016, https://ndpr.nd.edu/news/contemporary-dualism-a-defense/.

INDEX

Note: page numbers in *italics* refer to illustrations